Pests and Pathogens

plant responses to foliar attack

ENVIRONMENTAL PLANT BIOLOGY series

Editor: W.J. Davies
Institute of Environmental and Biological Sciences, Division of Biological Sciences, University of Lancaster, Lancaster LA1 4YQ, UK

Abscisic Acid: physiology and biochemistry

Carbon Partitioning: within and between organisms

Pests and Pathogens: plant responses to foliar attack

Forthcoming titles include:

Water Deficits: from cell to community

Pests and Pathogens

plant responses to foliar attack

P.G. Ayres
Institute of Environmental and Biological Sciences, Division of Biological Sciences, University of Lancaster, Lancaster LA1 4YQ, UK

βIOS
SCIENTIFIC
PUBLISHERS

First published in the United Kingdom 1992 by
BIOS Scientific Publishers Limited,
St Thomas House, Becket Street, Oxford OX1 1SJ, UK.

A CIP catalogue record for this book is available from the British Library.

ISBN 1 872748 01 5

Typeset by Alden Multimedia Ltd, Northampton, UK.
Printed by Information Press Ltd, Oxford, UK.

Preface

The study of plant responses to environmental stresses has been a botanical growth area in the last 20 years. Understanding has advanced rapidly not just because of the advent of the techniques of molecular biology but, also, because of the growing realization that from the cellular to the organismal level there is much similarity between plant responses to different stresses.

It is all the more unfortunate, therefore, that plant responses to pests and to pathogens have usually been studied in isolation. This is despite the facts that, for at least the last 20 years in most British universities, young scientists have been trained as biologists, rather than botanists (plant pathologists) or zoologists (entomologists), and that plants in the field are often assailed simultaneously by pests and pathogens. Both pests and pathogens are biotic stress factors, both are exploiters (a convenient shorthand for both groups) of the photosynthetic ability of plants and, as is now becoming increasingly recognized, both evoke similar growth and defence responses.

This book arises from a meeting on Plant Responses to Pests and Pathogens, held 9–10 April 1992 at the University of Lancaster, under the auspices of the Society for Experimental Biology (SEB). The object of the meeting was to explore the extent to which responses to pests and pathogens are the same and, thereby, to facilitate more cross-boundary thinking and progress in research. Contributors were selected either from that relatively small group already thinking about the similarity of responses or from the group whose expertise on one side of the boundary could usefully be extended to the other.

My thanks are due to the Series Editor, Bill Davies, for his advice and encouragement, the SEB for financial support, and the authors for their commitment.

Peter Ayres (*Lancaster*)

Contents

Contributors

Ayres, P.G. Institute of Environmental and Biological Sciences, Division of Biological Sciences, Lancaster University, Lancaster LA1 4YQ, UK

Bastiaans, L. Department of Theoretical Production Ecology, Wageningen Agricultural University, PO Box 430, 6700 AK Wageningen, The Netherlands

Biere, A. Netherlands Institute of Ecology, Department of Population Biology, Boterhoeksestraat 22, 6666 GA Heteren, The Netherlands

Bowles, D. Centre for Plant Biochemistry and Biotechnology, University of Leeds, Leeds LS2 9JT, UK

Edwards, P.J. Department of Biology, The University, Southampton SO9 5NH, UK

Esquerré Tugayé, M.T. Centre de Physiologie Végétale, Université Paul Sabatier, 118 route de Narbonne, 31062 Toulouse, France

Farrar, J.F. School of Biological Sciences, University of Wales, Bangor, Gwynedd, UK

Fournier, J. Centre de Physiologie Végétale, Université Paul Sabatier, 118 route de Narbonne, 31062 Toulouse, France

Linders, E.G.A. Netherlands Institute of Ecology, Department of Plant Population Biology, Boterhoeksestraat 22, 6666 GA Heteren, The Netherlands

de Nooij, M.P. Netherlands Institute of Ecology, Department of Plant Population Biology, Boterhoeksestraat 22, 6666 GA Heteren, The Netherlands

van Oijen, M. Centre for Agrobiological Research (CABO-DLO), PO Box 14, 6700 AA Wageningen, The Netherlands

Parker, M.A. Department of Biological Sciences, State University of New York, Binghampton, New York 13902–6000, USA

Paul, N.D. Institute of Environmental and Biological Sciences, Division of Biological Sciences, Lancaster University, Lancaster LA1 4YQ, UK

Prins, A.H. Department of Population Biology, University of Leiden, PO Box 9516, NL-2300 RA Leiden, The Netherlands

Rabbinge, R. Centre for Agrobiological Research (CABO-DLO), PO Box 14, 6700 AA Wageningen, The Netherlands

Rickauer, M. Centre de Physiologie Végétale, Université Paul Sabatier, 118 route de Narbonne, 31062 Toulouse, France

Rossing, W.A.H. Department of Theoretical Production Ecology, Wageningen Agricultural University, PO Box 430, 6700 AK Wageningen, The Netherlands

Scholes, J.D. Robert Hill Institute, Department of Animal and Plant Sciences, University of Sheffield, Sheffield S10 2TN, UK

Strange, R.N. Department of Biology, University College London, Gower Street, London WC1E 6BT, UK

Verkaar, H.J. Department of Population Biology, University of Leiden, PO Box 9516, NL-2300 RA Leiden, The Netherlands

van der Werf, W. Department of Theoretical Production Ecology, Wageningen Agricultural University, PO Box 430, 6700 AK Wageningen, The Netherlands

1

Plants versus pests and pathogens: an old story but the same story?

P.G. Ayres

1.1 Introduction

An injury or wound caused by a foliar pest or pathogen evokes a variety of responses which, for convenience, we regard as being related either to wound healing, prevention of further wounds, or minimizing the effects of the wound. For a plant there are no such clear cut distinctions. All responses require energy and carbon skeletons, and the different responses require co-ordination by shared regulatory mechanisms.

Recent findings, sometimes controversial, now justify us re-asking the old question, To what extent do plants have common responses to pests and pathogens? These findings relate to the nature of the molecules that elicit wound healing and defence reactions. The genes that control those responses are being identified, and exciting discoveries are being made about the long-distance signalling that occurs within and, even, between plants after wounding. Also, the nature and properties of the substances synthesized either *de novo* or in increased amounts after different wound events are becoming well characterized. So too are the nature and properties of substances selected for the constitutive defence of later generations. Finally, a complementarity is being revealed between our knowledge of the physiological responses of individual plants and our attempts to model mathematically the growth of populations in the field.

1.2 Common origins: rapid, cellular responses to wounding

The Angiosperms and Gymnosperms which dominate our modern flora have throughout their evolution had to cope with potential injury from pests and pathogens. Those biotic stresses were, however, episodic when compared with the

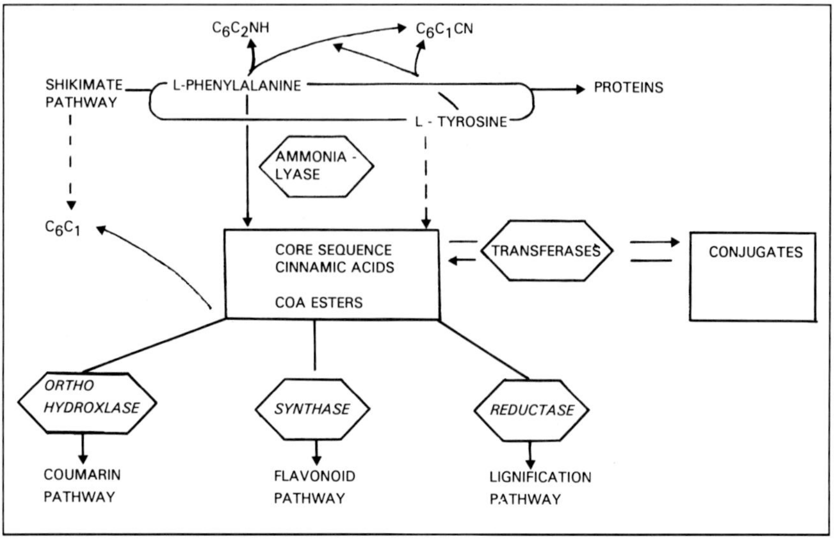

Figure 1.1. *Biosynthesis of phenolic products from shikimic acid. Phenylalanine and tyrosine are precursors in the synthesis of phenylpropanoids, from which are derived most common phenolics, e.g. hydroxycoumarins, lignans, lignins, flavonoids and isoflavonoids. See Figure 6.1 for further detail.*

more continuous abiotic stresses, for example, of a saline soil or a desiccating atmosphere, which challenged the first land plants. Thus, plants today possess a set of basic structures and processes that evolved primarily to enable them to cope with physical stresses, backed up by others which probably enabled appropriate responses to be made after the occasional failure, i.e. injury.

In the evolution of the phenylpropanoid pathway (Figure 1.1), which is the starting point for the synthesis of so many defence-related compounds, the initial driving force may have been the need to synthesize flavonoids to give protection from damaging ultraviolet light as plants moved onto land (Heath, 1991). Genes coding for phenylalanine ammonia lyase, like many coding for defence-related proteins, are developmentally regulated and are normally expressed at particular phases in the development of healthy plants (Liang *et al.*, 1989). Modifications to cell walls, such as lignification, or the deposition of hydoxyproline-rich glycoproteins, may have evolved to cope with mechanical rather than biotic stresses. Callose is an impermeable substance deposited rapidly wherever membranes or cells walls have to be sealed; such a response was probably needed in plant evolution to deal with mechanical wounding, rather than biotic wounding. Abiotic stresses continued to be a major driving force as plants became more complex. Rayner and Boddy (1988) suggest that the deposition of gums and resins in sapwood protects living trees from pathogens not directly, as was once thought, but indirectly; the barriers probably evolved to stop embolisms spreading through wounded xylem. By stopping air access to the tissues, they induce a lack of oxygen which, rather than any physical effect, inhibits fungal spread.

Recent advances in molecular genetics confirm that genes regulating basic developmental processes have been modified to control responses to biotic stresses. For example, in tobacco, two flower-specific acidic glucanase cDNAs were sequenced by Ori *et al.* (1990) who found the encoded enzymes were 80% similar to virus-induced acidic glucanases. As more is learned about the genes regulating resistance, so it becomes increasingly possible that they might be manipulated to man's advantage, e.g. so that their expression is intensified or prolonged. The first genetically modified plants that resist insect damage and viral infection are likely to become commercially available within the next 3 years (Braun *et al.*, 1991).

In attempting to understand whether plants use the same systems for recognizing and responding to pests and pathogens as for the basic event, wounding, a major impediment is our still restricted knowledge of the latter. Wounding drastically alters the plant's pattern of gene expression, both locally and systemically. Physical wounding leads to a loss of compartmentation, *inter alia* exposing the host cell wall to endogenous polygalacturonases. Additional polysaccharidases may be released by invading pathogens, or in the saliva of some herbivores, e.g. aphids. The oligosaccharides that are liberated in consequence are effective (endogenous) elicitors of gene expression but, because of their molecular size, their mobility is restricted. It is most unlikely they can directly alter gene expression at sites distant from the wound. For the same reasons of size, it is unlikely that oligosaccharides from microbial cell walls (exogenous elicitors) have systemic effects, although locally they may be extremely effective synergists for endogenous elicitors. Possibly, the synergy is an evolutionary refinement to ensure that only major wounds elicit major responses. Hartley and Lawton (1991) have recently speculated that the reason plants often respond more to insect feeding than to wounding (e.g. Kogan and Fischer, 1991) may be that insect mouthparts may be contaminated with microbes. Avoidance of the colonization of wounds by opportunist pathogens and, thus, the containment of injury must have been a potent force directing the evolution of defence responses.

Although they have received little attention in plant pathology, action potentials may prove to have a role in defence responses both locally and systemically (Davies, 1987). Oligosaccharides are among the substances that can initiate membrane depolarization, as described by Bowles in Chapter 3. Changes in membrane potential affect ion fluxes, in particular, fluxes of Ca^{2+} which in turn can affect the activity of many enzymes involved in cell wall metabolism, for example callose synthase. Ethylene has often been considered a candidate for the role of 'wound signal'. Despite much attention, confusion still exists about its exact role and it may be merely a symptom not a signal in the wound response. The potato and tomato proteinase inhibitors (e.g. *pin 2*) are amongst the best studied of gene families that are systemically induced. Data suggest that here ethylene may participate in a systemic pathway induced by wounding, but that ethylene is involved with genes other than *pin 2* (Sanchez-Serrano *et al.*, 1991). Perhaps not unsurprisingly, the conclusion must be that different genes are triggered by different signals.

Abscisic acid (ABA), unlike ethylene, is an effective messenger in the *pin 2* system, inducing *pin 2* mRNA in both ABA treated and untreated tissues (Sanchez-Serrano *et al.*, 1991). Endogenous ABA concentration rises 3- to 5-fold in wounded and distant tissues, although transport of ABA from the wounded to the distant reacting sites has

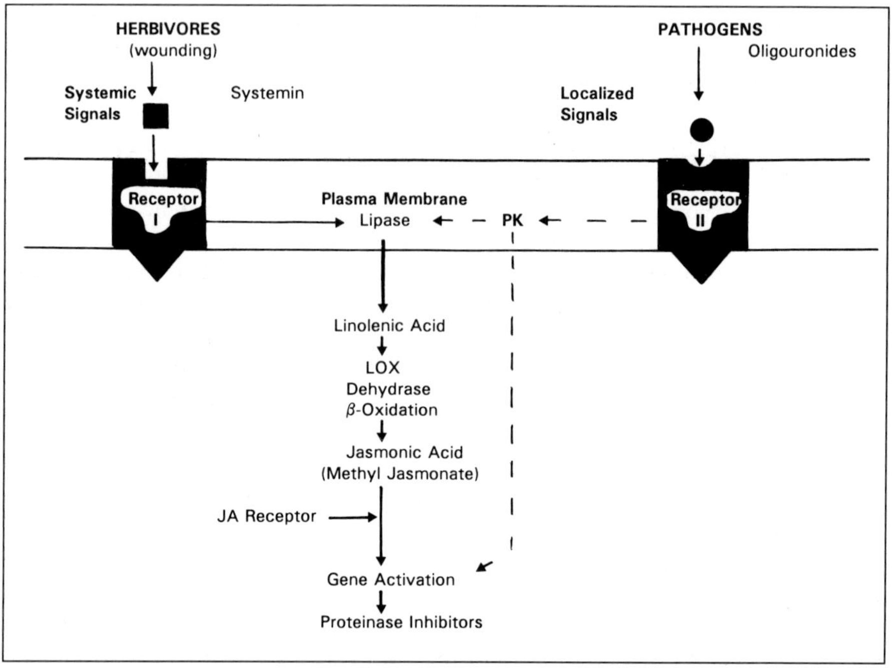

Figure 1.2. *Herbivores and pathogens may produce different signals but through a common pathway may cause the expression of wound-inducible proteinase inhibitors in tomato leaves. Modified from the model proposed by Farmer and Ryan (1992). PK = protein kinase; LOX = lipoxygenase; JA = jasmonic acid.*

yet to be demonstrated. Jasmonic acid (JA), which is produced from linolenic acid by the action of lipoxygenase, and which may derive ultimately from damaged membranes (Figure 1.2), also induces *pin 2* mRNA. It does this in ABA-deficient mutants, as well as in ABA-sufficient plants, suggesting that JA normally functions 'downstream' from ABA in the transduction pathway. At present there is insufficient evidence to distinguish between this possibility and the alternative that the steps requiring ABA and JA are located in different pathways. Farmer and Ryan (1992) have recently suggested that systemic signals arising from herbivore attack, e.g. systemin, an 18-amino acid polypeptide, which is a wound-induced phloem-mobile inducer of proteinase inhibitor in tomato (Pearce *et al.*, 1991), and local signals arising from pathogen attack, e.g. oligogalacturonides, may find a common mechanism of action through the activation of plasma membrane-bound lipase (Figure 1.2). Gabriel and Rolfe (1990) have suggested how different transmembrane proteins may fulfil the roles of receptors for different types of elicitor but, until such receptors are isolated and characterized, the relative importance of different signalling pathways remains uncertain.

Perhaps most remarkable of all, because of its ecological implications, Farmer and Ryan (1990) have demonstrated that proteinase induction by wounding can be transmitted between plants by the volatile messenger, methyl jasmonate, the product of de-esterification of JA. Methyl jasmonate in nanomolar quantities is effective, even in passing the message between plants of different species.

1.3 Constitutive defence: responses over generations

Defence depends on a battery of mechanisms, some constitutive ones being present before wounding or attack, although often enhanced afterwards, and other induced ones occurring only after the event. The same compounds may affect different groups of exploiter differently; conversely one exploiter may be affected by some defence compounds yet not by others. Thus, as well as being an anti-fungal agent, the soybean phytoalexin, glyceollin, is effective against oligophagous herbivores (Kogan and Fischer, 1991). Another feeding deterrent is salicylic acid (*o*-hydroxybenzoic acid), which is typical of simple phenolics present in higher concentrations after wounding than before (it is also another candidate for the messenger that transmits systemically acquired resistance, see Rasmussen *et al.*, 1991). However, while glyceollin did not deter feeding by the polyphagous soybean looper larva, *Pseudoplusia includens*, but reduced its growth rate, salicylic acid was a feeding deterrent but did not affect growth rate.

In considerations of plant defence, pathologists have emphasized the importance of induced mechanisms, while those studying herbivory (mostly entomologists) have emphasized the role of constitutive mechanisms. Further, most pathologists have studied crop plants, usually those where there is evidence of parasite-specific resistance (preferably of a gene-for-gene nature), because of their interest in the coupled phenomena of recognition and response at the cellular or, more recently, the molecular level. Comparatively little is known about non-host resistance to pathogens and, until very recent interest in weed control, wild species have been overlooked. It is important to know whether this difference between the scientific disciplines reflects a biological difference or merely their separate development.

Generally, pathogens injure tissue at a rate determined by their growth and do so markedly more slowly than herbivores. Exceptions may be those pathogens that release diffusible toxins, although even in their case some growth of the pathogen must occur before significant amounts of toxin are released, and those herbivores that suck sap or inject mobile toxins into sap. Another important difference between pathogens and herbivores is the greater mobility of the latter. Most herbivores can move rapidly within or between plants in response to visual or chemical stimuli. In contrast, it is only at the end of each life cycle (at least 7 days in most fungi) that significant movement of most pathogens can occur, even within a plant. Thus, although wound healing and the prevention of microbial invasion is essential following herbivory, and appropriate mechanisms would have been strongly selected through evolution, there would have been less pressure for the selection of locally induced defence mechanisms against the herbivores themselves. (The different importance attached to systemic protection against herbivores, and local defence against pathogens is reflected in the hypothesis outlined in Figure 1.2). This is not to say that systemic signals, or signals between plants, were not advantageous in defence against pathogens, and such mechanisms have evolved in several families (Dean and Kuc, 1987 ; Murray and Walters, 1992).

The extent to which levels of constitutive defence against herbivory are related either to the apparency of the plant or the richness of resources in its environment has been a matter of recent debate. The first theory was that apparent plants, e.g. those that are large or have a predictable growth habit, and which are often typical of climax vegetation, defended themselves with quantitative, typically highly reduced, energy-rich compounds of low toxicity, e.g. tannins or lignin, which were only overcome by specialist feeders. In contrast, unapparent plants, e.g. those that are small or ephemeral, were under less pressure from herbivores and could effectively defend themselves with qualitative compounds, typically toxic at low concentrations, e.g. alkaloids. Such a distinction is no longer accepted. The same substances, e.g. terpenes, may be growth regulators at very low concentrations in one situation and quantitative compounds, at high concentrations, in another. Quantitative compounds, e.g. polyphenols, may be oxidized as the result of herbivory, or penetration by pathogen, and thus serve as precursors for toxins.

Consideration of the carbon and nitrogen costs involved in the construction of specific compounds led to the second theory, which proposed that the highest levels of defence compounds are associated with species of resource-limited environments, possibly because the potential costs of replacing lost tissue are so great. Such defence compounds do not reduce plant vitality since they are formed at the expense of excess photosynthate. It is argued by Edwards, in Chapter 6, that neither of these theories is acceptable ; the *paramount* force driving the evolution of secondary metabolism has been the regulation of primary metabolism, not defence against herbivory. These secondary metabolites represent a resource upon which evolution has exerted selection. Such reasoning is clearly parallel to that of Heath (1991) regarding resistance to pathogens (see Section 1.2).

There is little evidence as to whether or not the slowest growing plants, typical of stressful environments, have relatively high levels of constitutive defence against pathogens. There was a widely promulgated notion that plants growing under sub-optimal conditions were relatively susceptible to disease, which is certainly untrue for biotrophic pathogens, where susceptibility is enhanced by the high light and nutrient conditions that lead to rapid growth. Paul (1990) contrasted the absence of rust and smuts on *Nardus stricta*, a dominant grass of nutrient-poor acid moorlands, with the presence of no less than three rusts and three smuts on *Agropyron repens,* a grass typical of richer sites. Although interesting, such slim evidence does not justify any far reaching conclusions; it merely reminds us that we are woefully ignorant of the incidence of pathogens on wild plants.

1.4 Slow growth responses of the whole plant

Even without intraplant spread of the injurious agent, an initially localized injury has repercussions that spread in time and space like ripples on a pond (Figure 1.3). Whereas an abiotic wound, or even a leaf-cutting ant, may simply remove photosynthetic tissue, a sap-sucking herbivore or a long-lived pathogen will continuously drain carbon, nitrogen and other nutrients from the host (see Box). The loss of dry matter, relative to healthy plants, may compromise a plant's ability to cope with environmental stress, particularly when growing naturally in mixture with healthy plants of the same or other species. Thus, a compromised plant may have a smaller, or shallower, root system, making it prone to drought, or it may have a shorter stem, causing it to be increasingly shaded by neighbours. If attack occurs early in the life of the plant, there may be opportunity costs in the sense that space may be conceded to neighbours, never to be regained (Gulmon and Mooney, 1986).

The magnitude of nutrient losses to pathogens and herbivores

No direct comparisons have been made by experiment but, excluding carbon losses through respiration of the exploiter organism, losses to aphids and biotrophic pathogens *at the time of peak exploitation* appear to be of a similar order. Thus, on 5-week-old barley plants, cv. Julia, of dry weight approx. 500 mg and infested for 4 weeks with bird cherry-oat aphid, *Rhopalosiphum padi*, biomass of the exploiter was 92 mg $plant^{-1}$ (Mallott and Davy, 1978). The exploiter biomass : plant dry weight ratio was 1 : 5.4 . Kneale and Farrar (1985) calculated that in barley, cv. Maris Mink, infected for 7 days with brown rust, *Puccinia hordei*, the amount of fungal material present was 260 $\mu g\ cm^{-2}$ leaf area. Thus, if we assume a value of 4 mg dry weight cm^{-2}, the exploiter biomass:plant ratio was 1:15. However, measurements were made at the beginning of spore production. If the ratio had been calculated 7 days later, after the peak of sporulation, it would have been much closer to unity since, from studies of *Puccinia recondita* on wheat over an equivalent period 7–14 days after infection, Mehta and Zadoks (1970) found that spore dry weight may be approximately equal to half the dry weight of the leaf.

Responses to herbivores and pathogens are, however, not merely confined to resisting an initial injury. Plants are able to compensate for injury by altering their physiology and morphology. The extent to which this can occur, i.e. the plant's plasticity, depends upon the severity of infection and the age of the plant when injured (see Chapter 2 by Prins and Verkaar). It will depend on the success of induced defence mechanisms in preventing repeated injuries, and will probably also depend upon the nature of the injurious agent. For example, an exploiter producing a mobile toxin, e.g. fusicoccin from *Fusicoccum amygdali* (Turner and Graniti, 1976), or sequestering significant amounts of solute(s) needed by distant organs, e.g. nitrate by powdery mildew of barley (Murray and Ayres, 1986), must limit a plant's capacity for compensation.

Compensation or recovery can take several forms. Increases in net photosynthetic rate can occur in healthy tissues of attacked plants. This increase may be in tissues surrounding the lesion or wound (Scholes, Chapter 7), but is often in young leaves that expand after the initial attack (Prins and Verkaar, Chapter 2). The photosynthetic capacity of the plant may be protected by changes in dry matter partitioning such that leaf area ratio or, sometimes, leaf area duration is increased. In response to herbivores

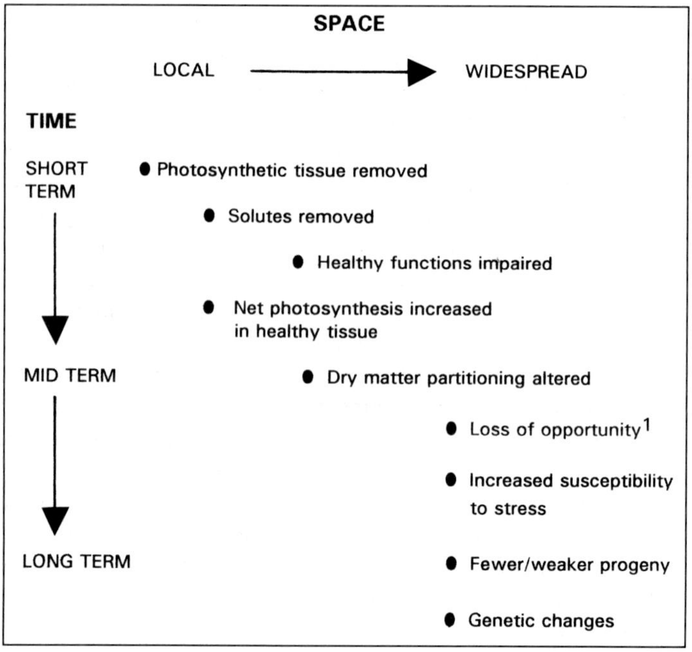

***Figure 1.3**. Generalized scheme to show the spread of impacts, in time and space, after localized foliar injury caused by a pest or pathogen.* [1]Sensu *Gulmon and Mooney (1986).*

and foliar pathogens, the shoot is typically protected at the expense of the root but, even below ground, compensatory increases in specific root activity in nutrient uptake may help to compensate for the absolute reduction in root dry matter (Paul, 1989; Ayres, 1991).

There is currently a gap, albeit narrowing, between our understanding of, on the one hand, the physiological processes involved in responses to pests and pathogens and, on the other, the growth of plant populations in the field. Under optimal conditions, the latter is directly proportional to the amount of solar radiation that is intercepted (RI), which is sometimes integrated to give healthy area absorption, HAA (Waggoner and Berger, 1987). The efficiency with which the trapped light energy is converted into chemical energy (RUE = radiation use efficiency) is relatively constant over periods of 14 days or more in healthy plants (Spitters, 1990), and determines the slope of the biomass/light interception relationship. Given that this relationship may have to be restricted to the period when the harvested portion of the crop is developing rapidly, because pests or pathogens may affect harvest index, it has been the basis of attempts to divide pests and pathogens into those that affect RI and those that affect both RI and RUE. For example, in three cereal diseases, glume blotch of wheat (*Septoria nodorum*), leaf blotch of barley (*Rhynchosporium secalis*), and brown rust of wheat (*Puccinia recondita*), effects on growth were explained solely by the reduction in RI that they caused (Van der Werf *et al.*, 1991), and a comparable relationship held for avocado infected with brown mite (Sances *et al.*, 1982). The analysis does have limitations, however, as explained by Rossing *et al.* in Chapter 11.

One limitation mentioned only briefly by Rossing is the complexity of biotic interactions in the field. In simulating damage to winter wheat caused by an aphid, *Sitobion avenae*, Wellings *et al.* (1989) found direct damage of 300 kg ha^{-1} was attributable to the withdrawal by aphids first of carbohydrates and later of nitrogen. From late August onwards, however, when the aphid population was collapsing, the major loss component was attributable to the honeydew blanket which covered leaves and intercepted radiation before it reached leaves (Figure 1.4). In an experimental study of sooty mould, *Capnodium* sp., growing on pecan foliage after aphid infestation and benefitting from the honeydew blanket, it was found that heavy fungal infection could reduce photosynthetically active radiation at the leaf surface by 98% (Wood *et al.*, 1988). Sooty moulds are, in general, pigmented fungi that grow on the leaf surface of senescing or aphid-infested leaves; their invasive ability is limited but they may reduce photosynthetic capacity to an extent yet unappreciated.

On physiological grounds, impairment of RUE might be expected where the pest or pathogen either produces a mobile toxin, or removes large amounts of assimilate, or causes a general reduction in turgor and leaf conductance. Damage caused by a leafhopper, *Emposasca fabae*, which injects a mobile toxin into the leaf while feeding, fits this bill (Johnson, 1987) Interestingly, damage to peanut infected by *Cercospora* leaf spot was also much greater than predicted from visible symptoms, suggesting reduced RUE. Presence of a mobile toxin is indicated, a possibility strengthened by

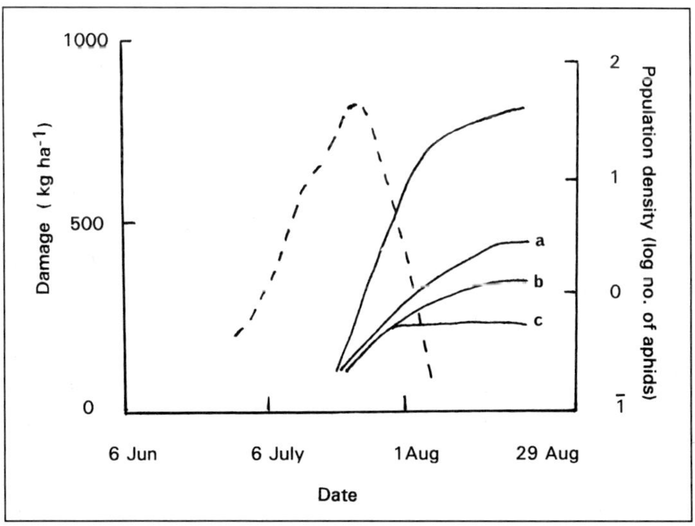

***Figure 1.4**. Simulation of the growth of an aphid,* Sitobion avenae, *on winter wheat (– – –) and total damage to the crop (——). Components of the total are (a) damage in the absence of honeydew effects on light-use efficiency, (b) direct aphid damage through withdrawal of both carbohydrates and nitrogen, and (c) direct damage through withdrawal of carbohydrate alone. Modified from Wellings* et al. *(1989).*

the knowledge that members of the genus *Cercospora* produce photosensitizing toxins (Daub and Briggs, 1983). Analysis of growth in the field might prove a stimulus to physiological investigations but, clearly, there is a need for closely integrated, complementary field and laboratory studies.

The initiation and expansion of leaves is remarkably sensitive to stress, whether by defoliation, reduced turgor, accelerated senescence or a multitude of other causes. Thus, it may be argued that, except in the short term, or in older plants where leaf expansion has ceased but photosynthesis is still active, effects of pests and pathogens on RUE will become rapidly translated into changes in RI or HAA. The significance of a change in RUE will be progressively diluted.

1.5 The future

Improved pest and disease resistance in domesticated plants is a continuing goal of plant breeding and engineering, and there is no shortage of new challenges as opportunities for the spread of pests and pathogens increase because of man's ever more rapid transport systems and, also, because of climate change. The production of plants resistant to multiple stresses is an attractive idea, but consideration must be given to the possible cost that resistance carries, particularly where it is constitutive. It has been reported that, in the absence of herbivores, pest-resistant varieties often have lower yields than susceptible varieties (Pimental, 1976), and the existence of comparable costs for resistance to pathogens has often been debated, e.g. Alexander (1990). However, as described by Parker in Chapter 12, there is as yet no compelling evidence that such costs are widespread where resistance to pathogens is concerned, in part because of the greater reliance that plants put on induced defence in their responses to pathogens.

Understanding has recently advanced rapidly not just because of the advent of the techniques of molecular biology but, also, because of the growing realization that from the cellular to the organismal level there is much commonality between plant responses to different stresses. Knowledge can be shared to mutual advantage as the evolutionary roots of those reponses are being laid bare. This is particularly true of responses to pests and pathogens which for too long have been considered in isolation from each other.

References

Alexander, H.M. (1990) Dynamics of plant-pathogen interactions in natural plant communities. In: *Pests, Pathogens and Plant Communities* (eds J.J. Burdon and S.R. Leather). Blackwell Scientific Publications, Oxford, pp. 31–48.

Ayres, P.G. (1991) Responses of root and shoot to pathogenic infection. In: *An Integrated Approach to the Study of Environmental Stress on Plant Growth* (eds H.A. Mooney, E.J. Pell and W.E. Winner). Academic Press, New York, pp. 227–248.

Braun, C.J., Jilka, J.M., Hemenway, C.L. and Turner, N.E. (1991) Interactions between plants, pathogens and insects: possibilities for engineering resistance. *Curr. Opin. Biotech.* **2**, 191–198.

Daub, M.E. and Briggs, S.P. (1983) Changes in tobacco cell membrane composition and structure caused by cercosporin. *Plant Physiol.* **71**, 763–766.

Davies, E. (1987) Action potentials as multifunctional signals in plants: a unifying hypothesis to explain apparently disparate wound signals. *Plant Cell Environ.* **10**, 623–631.

Dean, R.A. and Kuc, J.A. (1987) Immunization against disease: the plant fights back. In: *Fungal Infection of Plants* (eds G.F. Pegg and P.G. Ayres). Cambridge University Press, Cambridge, pp. 383–410.

Farmer, E.E. and Ryan, C.E. (1990) Interplant communication: airborne methyl jasmonate induces synthesis of proteinase inhibitors in plant leaves. *Proc. Nat. Acad. Sci., USA* **87**, 7713–7716.

Farmer, E.E. and Ryan, C.A. (1992) Octadecanoid precursors of jasmonic acid activate the synthesis of wound-inducible proteinase inhibitors. *Plant Cell* **4**, 129–134.

Gabriel, D.W. and Rolfe, B.G. (1990) Working models of specific recognition in plant-microbe interactions. *Ann. Rev. Phytopathol.* **28**, 365–391.

Gulmon, S.L. and Mooney, H.A. (1986) Economics of biotic interactions. In: *On the Economy of Plant Form and Function* (ed. T.J. Givnish). Cambridge University Press, Cambridge, pp. 667–698.

Hartley, S.E. and Lawton, J.H. (1991) Biochemical aspects and significance of the rapidly induced accumulation of phenolics in birch foliage. In: *Phytochemical Induction by Herbivores* (eds D.W. Tallamy and M.J. Raupp). John Wiley, Chichester, pp. 105–132.

Heath, M.C. (1991) Evolution of resistance to fungal parasitism in natural ecosystems. *New Phytol.* **119**, 331–343.

Johnson, K.B. (1987) Defoliation, disease and growth: a reply. *Phytopathology* **77**, 1495–1497.

Kneale, J. and Farrar, J.F. (1985) The localization and frequency of haustoria in colonies of brown rust on barley leaves. *New Phytol.* **101**, 495–505.

Kogan, M. and Fischer, D.C. (1991) Inducible defences in soybean against herbivorous insects. In: *Phytochemical Induction by Herbivores* (eds D.W. Tallamy and M. J. Raupp). John Wiley, Chichester, pp. 347–378.

Liang, X., Dron, M., Cramer, C.L., Dixon, R.A. and Lamb, C.J. (1989) Differential regulation of phenyl ammonia lyase genes during development and by environmental clues. *J. Biol. Chem.* **264**, 14486–14492.

Mallott, P.G. and Davy, A.J. (1978) Analysis of the effects of bird cherry-oat aphid on the growth of barley: unrestricted infestation. *New Phytol.* **80**, 209–218.

Mehta, Y.R. and Zadoks, J.C. (1970) Uredospore production and sporulation period of *Puccinia recondita* f. sp. *triticini* on primary leaves of wheat. *Neth. J. Plant Pathol.* **76**, 267–276.

Murray, A.J.S. and Ayres, P.G. (1986) Uptake and translocation of nitrogen by mildewed barley seedlings. *New Phytol.* **104**, 355–366.

Murray, D.C. and Walters, D.R. (1992) Increased photosynthesis and resistance to rust infection in upper, uninfected leaves of rusted broad bean (*Vicia faba* L.). *New Phytol.* **120**, 235–242.

Ori, N., Sessa, G., Lotan, T., Himmelhoch, S. and Fluhr, R. (1990) A major stylar matrix polypeptide (sp41) is a member of the pathogenesis related proteins superclass. *EMBO J.* **9**, 3429–3436 .

Paul, N.D. (1989) Effects of fungal pathogens on nitrogen, phosphorus and sulphur relations of individual plants and populations. In: *Nitrogen, Phosphorus and Sulphur Utilization in Fungi* (eds L. Boddy, R. Marchant and D.J. Read). Cambridge University Press, Cambridge, pp. 155–180.

Paul, N.D. (1990) Modification of the effects of plant pathogens by other components of natural ecosystems. In: *Pests, Pathogens and Plant Communities* (eds J.J. Burdon and S.R. Leather). Blackwell Scientific Publications, Oxford, pp. 81–96.

Pearce, G., Strydom, D., Johnson, S. and Ryan, C.A. (1991) A polypeptide from tomato leaves induces wound-inducible proteinase inhibitor proteins. *Science*, **253**, 895–898.

Pimental, D. (1976) World food crises: energy and pests. *Bull. Entomol. Soc. Am.* **22**, 20–26.

Rasmussen, J.B., Hammerschmidt, R. and Zook, M.N. (1991) Systemic induction of salicylic acid accumulation in cucumber after inoculation with *Pseudomonas syringae* pv. *syringae*. *Plant Physiol.* **97**, 1342–1347.

Rayner, A.D.M. and Boddy, L. (1988) *Fungal Decomposition of Wood.* John Wiley, Chichester.

Sances, F.V., Toscano, N.C., Hoffman, M.P., Lapre, L.F., Johnson, M.W. and Bailey, J.F. (1982) Physiological responses of avocado leaves to brown mite feeding injury. *Environ. Entomol.* **11**, 516–518.

Sanchez-Serrano, J.J., Amati, S., Ebneth, M., Hildmann, T., Mertens, R., Pena-Cortes, H., Prat, S. and Willmitzer, L. (1991) The involvement of ABA in wound responses of plants. In: *Abscisic Acid: Physiology and Biochemistry*. (eds W.J. Davies and H.G. Jones). BIOS Scientific Publishers, Oxford, 201–216.

Spitters, C.J.T. (1990) Crop growth models: their usefulness and limitations. *Acta Hortic.* **267**, 349–368.

Turner, N.C. and Graniti, A. (1976) Stomatal response of two almond varieties to fusicoccin. *Physiol. Plant Pathol.* **9**, 175–182.

Van der Werf, W., Bastiaans, L., Rossing, W.A.H. and Rabbinge, R. (1991) Modelling crop responses to growth reducing factors. In: *Biotic Stress of Barley in Arid and Semi-Arid Environments* (ed. J. Reisselman). Montana State University, Bozeman, pp. 87–102.

Waggoner, P.E. and Berger, R.D. (1987) Defoliation, disease and growth. *Phytopathol.* **77**, 393–398.

Wellings, P.W., Ward, S.A. Dixon, A.F.G. and Rabbinge, R. (1989) Crop loss assessment. In: *Aphids Their Biology, Natural Enemies and Control* Volume C (eds A.K. Minks and P. Harrewijn). Elsevier, Amsterdam, pp. 49–64.

Wood, B.W., Tedders, W.L. and Reilly, C.C. (1988) Sooty mold fungus on pecan foliage suppresses light penetration and net photosynthesis. *Hortic. Sci.* **23**, 851–853.

2

Defoliation: do physiological and morphological responses lead to (over)compensation?

A.H. Prins and H.J. Verkaar

2.1 Introduction

To a plant, defoliation is the most striking aspect of herbivory. It has immediate, negative effects on plant performance and, as a consequence, it probably affects growth, reproduction and fitness in the long term. However, one of the most puzzling discoveries made by plant ecologists and evolutionary biologists in the last decade is that an increase in plant growth and fitness sometimes results from herbivory. The consumption of vital plant organs sometimes actually seems to increase plant fitness! What constraints are released by herbivory, why do unattacked plants seem to grow or reproduce suboptimally? In this chapter, we try to answer these questions by looking in more detail at various responses to herbivory.

Plants exhibit a range of mechanisms in their responses to herbivory, the effects of which can be conveniently divided into physiological and morphological (Table 2.1). Physiological mechanisms include a stimulation of photosynthetic rate, a changed hormonal status, a redistribution of carbohydrates and mineral nutrients, and delayed plant senescence. The morphological mechanisms include changes in leaf area ratio, in plant architecture and in root development. We discuss here whether these mechanisms, either individually or collectively, can compensate partly, or even fully for tissue damage.

Plant compensation must be seen in the context of the biotic and abiotic environment: it is also affected by the plant growth stage during which damage takes place, the time available for recovery, grazing intensity and frequency. Competition and abiotic conditions affect the potential for compensation. Plants with different life strategies may show specific reactions to herbivory. These modifying factors will be discussed after we have described the possible reactions to herbivory.

Table 2.1. *Plant responses to defoliation/herbivory*

	Comments	Species	Lab (L)/field (F)	References
PHYSIOLOGICAL MECHANISMS				
Increased NAR	–	*Pinus radiata*	L	Sweet and Wareing (1966)
	–	*Nicotiana tabacum*	L	Kolodny-Hirsch *et al.* (1986)
	Depends on N-availability	*Cynoglossum officinale* *Verbascum thapsus*	L	Verkaar *et al.* (1986)
	Limited duration	*Cynoglossum officinale* *Senecio jacobaea*	L	Prins *et al.* (1989)
	Limited duration	*Briza subaristata*	L	Oesterheld and McNaughton (1991)
	For a longer period, due to repeated clipping	*Bouteloua curtipendula*	L	Ingham and Detling (1986)
Reduced NAR	At site of infection, temporary increase in younger leaves	Several species	L	Ayres (1991)
	–	*Brassica* spp.	L	Van Emden (1990)
	–	*Lycopersicon*	L	Royalty and Perring (1989)
Increased carbohydrate investment in above-ground plant parts	Total biomass yield reduced by defoliation	*Agropyron smithii*	L	Polley and Detling (1989)
	Total biomass yield reduced by defoliation	*Bouteloua gracilis*	L	Jaramillo and Detling (1988)
	Rapidly forming new leaves	*Euphorbia esula*	F	Cyr and Bewley (1990)
	Remobilization proportional to initial contents and intensity of herbivory	*Trifolium repens*	L	Baur-Höch *et al.* (1990)
Increased nitrogen investment in above-ground plant parts	Mobilized as soluble protein to newly formed leaves	*Trifolium subterraneum*	L	Culvenor *et al.* (1989) Culvenor and Simpson (1991)
		Trifolium repens	L	Marriott and Haystead (1990)

Delayed plant senescence	Inhibition of protein, chlorophyll and enzyme degradation, reduced starch accumulation, increased retention of mature leaves	*Nicotiana tabacum*	F	Kolodny-Hirsch *et al.* (1986)
	Regulated by hormonal activity	Several plant species	L	Thimann (1980, 1987) Nooden (1980)
	Only for two oldest, fully expanded leaves, senescence not affected in all other foliage elements	*Agropyron desertorum* *Agropyron spicatum*	F	Nowak and Caldwell (1984)
MORPHOLOGICAL MECHANISMS				
Increase in LAR	Increase higher in severely defoliated plants	*Cynoglossum officinale* *Senecio jacobaea*	L	Prins *et al.* (1989)
		Senecio vulgaris	L	Paul and Ayres (1986)
Decrease in LAR		*Vicia faba*	L	Paul and Ayres (1986)
Stimulated growth from axillary buds	More inflorescences	*Jurinea mollis*	L	Inouye (1982)
	More inflorescences	*Senecio jacobaea*	F	Islam and Crawley (1983)
	Positive effect on seed production	*Ipomopsis aggregate*	F	Paige and Whitham (1987)
	Positive effect on seed production under some circumstances	*Ipomopsis arizonica*	F	Maschinsky and Whitham (1989)
	Negative effect on seed production	*Epilobium latifolium*	F	Doak (1991)
Changed leaf demography	Decreased leaf initiation and leaf survival	*Cardamine cordifolia*	F	Louda (1984)
Pseudogall formation	Induced by aphids	*Prunus padus*	F	Leather (1988)

Table 2.1. *(Continued)*

	Comments	Species	Lab (L)/field (F)	References
Changed root development	Decreased root length growth and root weight	*Agropyron desertorum*	F	Richards (1984)
	Decrease in root:shoot ratio	*Rumex crispus*	L	Bentley and Whittaker (1979)
	Decrease in root:shoot ratio	*Eucalyptus blakelyi*	L	Vranjic and Gullen (1990)
	Increase in root:shoot ratio	*Rumex obtusifolius*	L	Bentley and Whittaker (1979)
	Compensation for loss of above-ground tissue, at the expense of root growth	*Plantago lanceolata*	L	Fajer *et al.* (1991)
	Decrease in root:shoot ratio	Several species infected by pathogens	L	Ayres (1991)

2.2 Physiological reactions to herbivory

2.2.1 *Changes in efficiency of dry matter assimilation*

Defoliation imparts a severe physiological stress on plants by suddenly restricting the availability of photosynthate needed for growth. Young plants, and plants that have not yet developed a storage reserve system, suffer most from this sudden stress (Cyr and Bewley, 1990). Plants may be stimulated to compensate physiologically for the removal of tissue.

In a classical experiment with *Pinus radiata*, Sweet and Wareing (1966) showed that, 6 days after removing one-third or all the fully expanded leaves, defoliated plants had a higher net photosynthetic rate per unit of remaining leaf than undefoliated plants. The severely defoliated plants even had a higher photosynthetic rate than moderately defoliated plants. Periods of increased efficiency in assimilating dry matter were found after artificial damage in *Nicotiana tabacum* (Kolodny-Hirsch *et al.*, 1986). These correlated with increases in chlorophyll, soluble proteins, and ribulose-1,5-bisphosphate carboxylase oxygenase content. Such physiological responses are found fairly often in plants affected by defoliation (Alcock, 1964; Wallace *et al.*, 1984; Danell *et al.*, 1985) and have been noted, also, in upper healthy leaves of plants infected on lower leaves by biotrophic pathogens (Walters and Ayres, 1983). Pathogens, such as rust or powdery mildew fungi, were found to reduce the rate of net photosynthesis at the site of infection in groundsel, pea and barley, but in the younger leaves net photosynthesis was temporarily increased (Ayres, 1991).

An enhanced net assimilation rate (NAR) may lead to the successful replacement of leaf tissue. However, this response depends on nitrogen availability, e.g. in situations with a high nitrogen availability NAR was significantly higher in defoliated *Cynoglossum officinale* and *Verbascum thapsus* than in undefoliated plants, but this increase did not occur when nitrogen was not available (Verkaar *et al.*, 1986).

In *Senecio jacobaea* and *C. officinale*, the increase in NAR as a result of damage was only temporary (Figure 2.1). The higher NAR was found for a short period after defoliation and only at the higher level of defoliation. After moderate defoliation in *C. officinale* no increase in NAR was found in the first few weeks, but during the second part of the experiment NAR temporarily decreased less quickly than in the undefoliated plants (Figure 2.1) (Prins *et al.*, 1989). A significantly higher NAR may be found only for a short period after defoliation, the rate soon decreasing to equal that found in non-defoliated plants. This was also found after defoliation of the grass, *Briza subaristata*, a native perennial grass from the flooding pampas of Argentina (Oesterheld and McNaughton, 1991a).

Repeated clipping may prolong increased net photosynthesis: in *Bouteloua curtipendula*, defoliated at weekly intervals, rates of photosynthesis and transpiration per unit of leaf area at final harvest (7–95 days after planting) were two to three times higher in defoliated plants than in undefoliated plants (Ingham and Detling, 1986).

Defoliation causes a sudden change in carbon status and therefore in source/sink relationships, even within the shoot. The increase in NAR in the remaining leaves suggests that photosynthesis can be temporarily increased to meet an increased

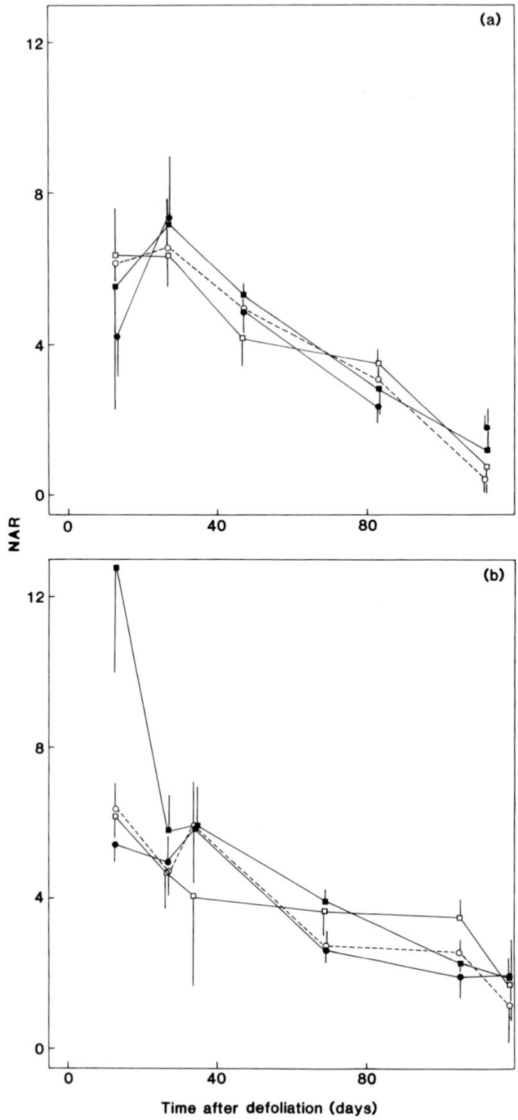

Figure 2.1*. Net assimilation rate (NAR) ($g\,g^{-1}\,day^{-1} \times 10^{-4}$) in (a)* Cynoglossum officinale *and (b)* Senecio jacobaea *after defoliation. (○) Control, (●) 10–12% defoliation, (□) 30% defoliation, (■) 94.5–98% defoliation. Standard errors are indicated to one side of the mean. From Prins* et al. *(1989).*

demand for photosynthates in the remaining root and the newly formed shoots (Neales and Incoll, 1968; Wareing and Patrick, 1975). However, no net increase and even decreasing levels of assimilation were found in some studies. The aphid, *Brevicoryne brassicae*, feeding on Brussels sprout, not only induced a decrease in the number of leaves and the leaf area, but the remaining leaves on the infested plants were less efficient assimilators. Aphid feeding results in leaf curling and a decreased density of stomata in new leaves, factors which might contribute to the difficulties the plant has in compensating for tissue damage (van Emden, 1990). Feeding of mites, *Aculops lycopersici*, caused a reduction in net photosynthesis of affected tomato leaflets. Net photosynthesis and mite-days per cm^2 of leaflet were significantly negatively correlated. Moreover, healthy tomato leaflets did not show an increase in photosynthesis to compensate for the reduction in the damaged leaflets (Royalty and Perring, 1989).

2.2.2 *Redistribution of carbohydrates and mineral nutrients*

Regrowth following herbivory can be increased by changes in the translocation pattern of newly formed carbon or by the redistribution of previously accumulated carbon. Translocation of carbohydrates and nitrogen compounds from the roots to the leaves can occur at rates higher than in ungrazed conditions. Changes in the steady-state processes of carbon fixation and translocation have been extensively studied, but, at this moment, the exact mechanism governing the changes following herbivory is unknown (Dyer *et al.*, 1991). Note that Farrar (in Chapter 8) proposes that no new mechanisms need to be involved to explain carbon partitioning in the potentially more complex situation where plants are infected by a microbe; the principles of turgor-driven bulk flow apply equally well to both healthy and infected plants.

In the grasses, *Agropyron smithii* and *Bouteloua gracilis*, an increased investment in above-ground relative to below-ground biomass and nitrogen was found after defoliation (Jaramillo and Detling, 1988; Polley and Detling, 1989). However, total biomass yield was significantly reduced. Following defoliation, certain herbaceous perennials redistribute carbon primarily from leaf and stem base to active meristems, thus rapidly forming new leaves (Cyr and Bewley, 1990). In stolons of *Trifolium repens*, carbohydrate reserves, mostly starch, were remobilized in proportion to their initial contents, and when plants were more severely damaged, remobilization was greater (Baur-Hoch *et al.*, 1990).

Remobilization of nitrogen following herbivore attack has received much less attention than carbohydrate remobilization (Culvenor and Simpson, 1991), probably because net changes are smaller (Smith and Silva, 1969; Caldwell *et al.*, 1981). In *Trifolium subterraneum*, 17% of the nitrogen present in storage organs (branch, taproot, fine roots, nodules) was mobilized as soluble protein to the newly formed leaves after defoliation (Culvenor *et al.*, 1989; Culvenor and Simpson, 1991). Comparable results were found for *T. repens* (Marriot and Haystead, 1990).

2.2.3 *Delayed plant senescence*

Following herbivory, the enhanced demand for assimilates from the remaining plant tissue may delay the normal decline of photosynthetic rate with age in the unattacked or partly defoliated leaves. Arrested senescence in response to leaf loss has been reported for a wide range of plants (Woolhouse, 1967; Wareing *et al.*, 1968). Regreening of yellow senescent leaves on vegetative plants of *Nicotiana tabacum* could be induced by removal of the younger leaves of the plant (Woolhouse, 1967). Not only leaf herbivory, but also the removal of reproductive organs, such as developing fruits or seeds, can delay the onset of senescence.

A delay in leaf senescence is characterized by the inhibition of protein, chlorophyll and enzyme degradation, reduced starch accumulation, and increased retention of mature leaves (Kolodny-Hirsch *et al.*, 1986). The control of senescence processes is regulated by hormonal activity (e.g. Nooden, 1980; Thimann, 1980, 1987).

A delay in senescence is not always documented for all leaves following herbivory: in *Agropyron desertorum* and *A. spicatum* the onset of senescence was found only for the two oldest, fully expanded leaves that were present when the plants were initially clipped; the initiation and senescence of all other foliage elements were not affected by the clipping treatments (Nowak and Caldwell, 1984).

2.3 Morphological reactions to herbivory

2.3.1 *Changes in leaf area ratio*

The morphological mechanism behind regrowth may involve an increase in leaf area ratio (LAR, $cm^2 g^{-1}$). In the biennials, *S. jacobaea* and *C. officinale*, LAR increased at moderate (30%) and even more at high (94–100%) degrees of defoliation (Figure 2.2) (Prins *et al.*, 1989).

In plants infected by pathogens, the inhibition of net photosynthesis in the infected leaves, as mentioned in Section 2.2.1, may be compensated by an increase in leaf area ratio (Paul and Ayres, 1986; Ayres, 1991). In nutrient-rich as well as nutrient-poor situations, an increase in LAR was found in groundsel, *Senecio vulgaris*. However, the opposite reaction has also been reported: *Vicia faba* infected by rust exhibited a decreased leaf area ratio (Williams, 1978).

2.3.2 *Changes in plant architecture*

Damage to the shoot apex by herbivory affects apical dominance and stimulates growth of branches from axillary buds, previously inhibited by auxins produced in the shoot apex (Salisbury and Ross, 1985). The release of competition for nutrients and carbohydrates between the axillary buds and the shoot apex may also contribute to the cessation of the inhibition of axillary buds (Benner, 1988).

An alteration in branching pattern following apex removal is frequently found (e.g. Taylor and Bardner, 1968; Inouye, 1982; Islam and Crawley, 1983; Paige and Whitham, 1987; Maschinsky and Whitham, 1989). Apex removal may induce plants that normally form a terminal inflorescence to produce a greater number of inflorescences. Nothwithstanding this greater number of inflorescences, a negative effect of herbivory on seed production is often found (e.g. Doak, 1991). This may be caused by the direct loss of tissue or by the loss of the future photosynthetic capacity of that tissue. Moreover, apex removal causes a delay in plant growth because the axillary branches first have to form leaves before the photosynthetic rate can reach the same level as before defoliation.

Various morphological reactions to herbivory are found when plant parts other than the apex are affected. In the birches, *Betula pendula* and *Betula pubescens*, a decrease in the number of long shoots was found, but the frequency of branched shoots

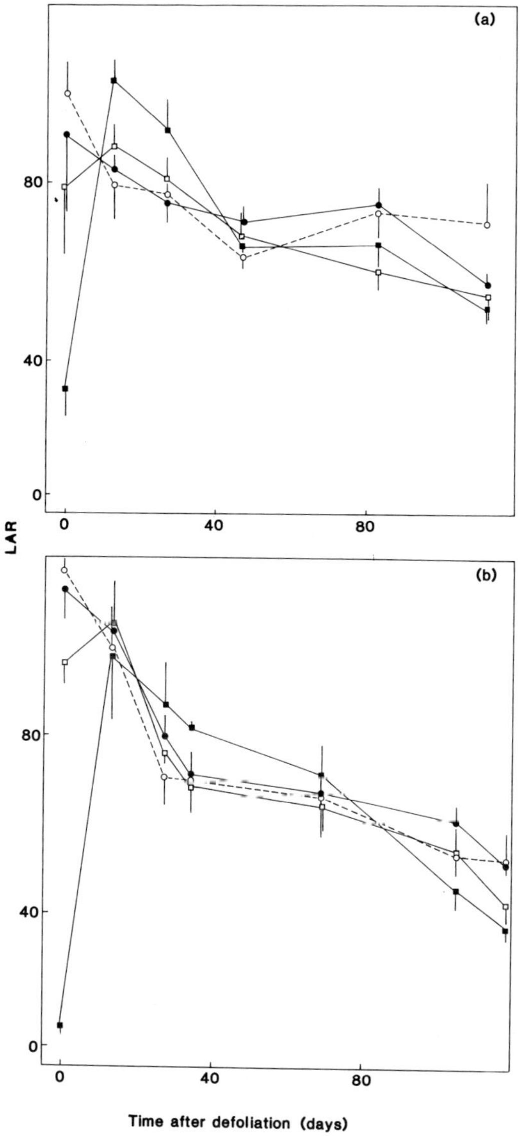

Figure 2.2. *Leaf area ratio (LAR, cm^2 g^{-1}) after defoliation in (a)* Cynoglossum officinale *and (b)* Senecio jacobaea. *(○) Control, (●) 10–12% defoliation, (□) 30% defoliation, (■) 94.5–98% defoliation. Standard errors are indicated to one side of the mean. From Prins* et al. *(1989).*

increased, as well as the length and dry weight of individual shoots (Danell *et al.*, 1985; Bergström and Dannell, 1987). Feeding by the weevil, *Unapsis euonymi*, increased leaf abscission in *Euonymus fortunei*. This may not have been detrimental to the plant because photosynthetic output of the leaves had already been impaired as a result of chlorosis in the insect-damaged leaves (Cockfield and Potter, 1986).

In the herbaceous crucifer, bittercress, *Cardamine cordifolia*, leaf herbivory well below complete defoliation (25% of the leaf area) changed leaf demography (leaf initiation and leaf survival) and reduced plant performance (a decrease in plant stature, leaf area and fruit production)(Louda, 1984).

Different herbivores may attack the same plant in different ways and, therefore, cause different morphological responses. Marked changes in plant architecture are found as a result of aphid infestation (*Rhopalosiphum padi*) on bird cherry (*Prunus padus*). Aphids induce the formation of a pseudogall at high infestation levels and the shoot tip is killed, preventing terminal bud formation. Attacked leaves are likely to reduce the photosynthetic capacity of the plant during the rest of the year, because attacked leaves are discoloured and malformed. In contrast, a moth, *Yponomeuta evonymellus*, defoliates the tree after bud formation is completed. The tree can form regrowth leaves after defoliation has ended (Leather, 1986; 1988).

2.3.3 *Leaf damage and root development*

The allocation of more resources to above-ground regrowth is a common and widespread reaction of plants to herbivory, as well as to infection by fungi (Ayres, 1991). Sometimes this is attended by a decrease in root-length growth and weight. In *Agropyron desertorum*, a grazing-tolerant bunchgrass, root-length growth of clipped plants was 50% less than that of intact plants. Relatively more resources were allocated to above-ground plant parts, so re-establishing the root:shoot balance. In a grazing-sensitive bunchgrass, *Agropyron spicatum*, root elongation was not affected by clipping. In this species, foliage replacement was prevented, and thereby the re-establishment of the root:shoot ratio. Herbivory may, thus, weaken the competitive position of *A. spicatum*, whereas the grazing-tolerant *A. desertorum* may maintain its competitive position (Richards, 1984). In the grazing-tolerant grass, *Sporobolus kentrophyllus*, a similar decrease in biomass allocation to the roots was found. Tillering rate increased in this species (Ruess, 1988).

Rumex obtusifolius responded to heavy grazing by an increase in the root:shoot ratio. *Rumex crispus*, on the other hand, responded to defoliation by a decrease of root:shoot ratio (Bentley and Whittaker, 1979).

Patterns of dry matter allocation were altered by infestation of *Eucalyptus blakelyi* by the sap-sucking herbivore, *Eriococcus coriaceus*. Proportionally more dry matter was allocated to the leaves at the expense of root growth. This led to a significantly lower root:shoot ratio in infested than in control plants (0.50 ± 0.25 *vs.* 0.79 ± 0.22; $P < 0.001$) (Vranjic and Gullan, 1990). In response to simulated herbivory, both in ambient and enriched CO_2 environments, compensation for the loss of above-ground tissue was found in the perennial herb, *Plantago lanceolata* (Fajer *et al.*, 1991), but at the expense of root growth. Although levels of reproductive output were maintained, negative effects of defoliation on growth, or increased mortality, could be found in later years.

These studies indicate that plants tend to re-establish their root:shoot balance after defoliation, as was proposed by Brouwer (1962a, b, 1963, 1983) in his theory on the functional equilibrium between overground and underground parts. As soon as this species-specific balance is reached, investment in above-ground plant parts may stop, and other mechanisms may contribute to compensate for the consequences of leaf losses.

The root:shoot ratio may also be altered as a result of a disease: for example, oat and barley infected with powdery mildew exhibited a decreased root:shoot ratio (Ayres, 1991). The causes of these changes may differ between species and are discussed in more detail by Farrar, in Chapter 8.

2.4 Compensation

2.4.1 *Timing of damage and time for recovery*

The mechanisms described above may partly or fully compensate for the negative effects of defoliation. Compensation is concluded when the cumulative dry weight of the treated plants equals that of the controls, whereas partial compensation occurs when the cumulative dry weight of the treated plants is less than that of the controls (Belsky, 1986). In terms of (over)compensation strategies, the effectiveness of the individual plant's reactions to defoliation strongly depends on growth stage, season, and environmental factors. Grazing systems are characterized by some pattern of independence from these factors, which may be quite persistent.

The possibilities for compensation are highly influenced by the developmental stage of the plant at the time of defoliation. Moreover, the moment during the season when a plant is affected by herbivores also influences possible compensation. These two factors may often be associated, especially in natural situations.

Simulated herbivory by Colorado beetle on potato cultivars resulted in greatest yield reductions when defoliation took place during the main flowering period. When herbivory was simulated through earlier or later defoliation, yield reductions were less, demonstrating that compensation is affected by the time of defoliation (Shields and Wyman, 1984). Artificial damage to *Salix smithiana* growing in central Sweden showed that the most severe growth reductions were when defoliation took place in early-mid summer. At that time, the most productive cohort of leaves was affected (Larsson, 1983). In *Agropyron desertorum*, tiller replacement was not affected by heavy grazing when this took place before the elongation of the internode. Defoliation after the elongation of the internodes (which elevates apical meristems) reduced the number and heights of replacement tillers and, moreover, increased the winter mortality of those tillers produced in autumn (Olson and Richards, 1988a). Turtlehead plants, *Chelone* spp., herbaceous perennials from wet meadows and fens, reacted to herbivory prior to flower bud development by a reduction in the number of reproductive stalks, flower buds, flowers and seed capsules. When plants were severely defoliated after flower buds appeared, only the final number of seed capsules and seeds per capsule were affected (Stamp, 1984). In the herbaceous crop, 'spring cabbage', *Brassica oleracea*, sensitivity to defoliation increased during the period from planting to head-formation. Sensitivity was maximal when plant leaf area reached approximately 2000 cm^2, and did not increase afterwards (Wit, 1985). All these studies indicate the importance of the timing of herbivory. Plants defoliated during flowering, or during the most productive period in the growing season, are more affected than those defoliated at other times.

Ability to recover from defoliation decreases during the growing season because progressively less time is available to overcome the delay in phenology (Inouye, 1982; Solomon, 1983; Maschinsky and Whitham, 1989). The physiological constraints on a plant may limit compensation of late damage even when environmental conditions are favourable. Therefore, damage later in the season may have an immediate effect on monocarpic species, for example resulting in a delay of flowering (Harris *et al.*, 1978). Theoretically, perennial plants may compensate in a later year for earlier damage.

Plants that flower later than normal may risk less than normal pollination. Thus, self-incompatible species may suffer most, but even in self-compatible species, e.g. *C. officinale*, individuals that flower later in the year produce significantly less seed (de Jong and Klinkhamer, 1991). This last effect emphasizes the importance of male fitness. Not only growth and seed production, but also possible effects on pollination, must be considered when studying a plant's reaction to herbivory.

2.4.2 *Grazing intensity*

Our knowledge of the effect of herbivory was, until the last decade, based on agricultural and forestry management practices. Establishing relationships between leaf area loss and yield loss gave an indication of the economic injury level. Correlating the consumption of a single herbivore with the maximum leaf area loss that could be sustained without any significant loss of final yield established the number of herbivores that could be safely allowed per unit leaf area. Growth and seed production were therefore the most important parameters studied. In ecological studies, however, implications for plant fitness are much more important.

In many species, low levels of herbivory do not lead to any reduction in final yield (Seastedt *et al.*, 1983; Olson and Richards, 1988b; Reichman and Smith, 1991). In many situations, growth is limited by environmental factors, and a little herbivory has no additional effect. In laboratory experiments, on the other hand, optimal conditions may quickly result in compensation for the negative effects of herbivory. Thus, when *S. jacobaea* and *C. officinale* suffered low (10%) or medium (30%) defoliation in the laboratory, plants took only 2 weeks to attain the same total plant biomass as undefoliated (control) plants. Complete defoliation affected plant dry weight for a longer period (Prins *et al.*, 1989). Plants with higher levels of defoliation require longer periods to recover than plants defoliated to a lesser extent. As long as the environmental conditions are favourable, however, even severely defoliated plants may recover, if given sufficient time.

2.4.3 *The plant's environment*

Most studies on compensation have been under controlled conditions. They are likely to overestimate greatly the ability of plants to compensate for herbivory. While under controlled conditions a shift from root to shoot allocation in defoliated plants may have little effect on nutrient or water uptake, the effects under changing natural conditions may be much larger. There is good evidence that the effects of herbivory are strongest under conditions of water stress, nutrient stress or competition.

Drought can interact negatively with plant compensation characteristics. For the biennial, *S. jacobaea*, decreasing moisture availability limited plant compensation. One irrigation after defoliation was sufficient to increase significantly the mean number of nodes per plant when compared with defoliated plants that were not supplied with water. Incidentally, these results are important for the study of biocontrol agents for *S. jacobaea* in areas where the plant has become a range weed; the full potential of herbivorous agents might be found in years with low moisture conditions (Cox and McEvoy, 1983). Water stress significantly reduced regrowth after clipping in the grass, *Hyparrhenia filipendula*. Frequent watering stimulated leaf elongation but had no impact on tillering. Nitrogen addition, on the other hand, had a strong impact on tiller number but not on leaf elongation, indicating that each form of stress can act differently on growth (Coughenor *et al.*, 1985).

Differences in nitrogen availability strongly influenced the regrowth capacity of *Verbascum thapsus* and *C. officinale* in the laboratory. Without nitrogen, *V. thapsus* died after defoliation. However, if given nitrogen, growth of defoliated and unclipped plants was similar. In *C. officinale*, defoliated plants that were not supplied with nitrogen did not die, but showed only very little regrowth (Verkaar *et al.*, 1986).

Simultaneous grazing by herbivores and other forms of stress, like competition, may have additive effects on plant performance (Harris, 1973; Whittaker, 1979; Cottam *et al.*, 1986). Pathogens, in combination with other forms of stress, were also found to have additive negative effects on plants (Ayres, 1991). In a situation of combined stress, opposed allocation responses may be found in individual plants. The other stress may impair those morphological and physiological responses to grazing that normally minimize the effects of herbivory on growth rate. Oesterheld and McNaughton (1991b) describe such a situation for the combined effect of grazing and flooding on Serengeti grasses. Grazing may induce plants to allocate resources to leaf production, whereas flooding may increase the allocation of resources to the roots, because the maintenance of roots in an anoxic environment may be costly. A synergism between invertebrate herbivory and another form of stress, i.e. competition, was found in *R. obtusifolius*. Only in competing plants did herbivory cause a significant reduction in plant growth (Cottam *et al.*, 1986).

Particular situations are conceivable, however, in which some forms of stress may interact positively with grazing. The removal of self-shading by grazing is an example. Under conditions of high atmospheric humidity and ample water supply, grazing may be beneficial by removing transpiring leaf area. Grazed plants may survive a drought period, whereas ungrazed plants may desiccate (Verkaar, 1988). Thus, in the grasses, *Hyparrhenia rufa* and *Trachypogon plumosus*, grazing and water stress interacted strongly; the effects of defoliation were reduced under water stress, probably because there was a more gradual depletion of soil water when the transpiring surface was reduced by grazing (Simoes and Baruch, 1991).

An unexpected interaction between herbivory and field conditions was found by Michaud (1989); herbivory led to reduced nectar production in flowers of fireweed, *Epilobium angustifolium*. Plants were thus less attractive to pollinators, resulting in reduced female and male fitness. As already mentioned, for a better understanding of the effects of herbivory on plant fitness, field studies are preferable, and both male and female fitness must be examined, a point made again by Parker in Chapter 12. Few studies have looked at effects on female fitness, and not a single study has examined male fitness. To understand compensation in an evolutionary context, the latter is by far the most important issue that should be addressed in herbivory experiments.

2.5 Overcompensation

During the last two decades, debate about the beneficial effects of herbivores has engaged plant ecologists. Although compensatory productivity, enabling plants to replace tissue lost to herbivores, was discussed by Eaton in 1931, the possibility of overcompensation was not considered until the much cited paper of Owen and Wiegert (1976). In this theoretical study, they proposed that an extension of the period of leaf fall, through the effects of primary consumers, might increase decomposition and, therefore, nutrient conservation and utilization efficiency, thus resulting in an overall increase in plant fitness. However, there has been little evidence until recently that plants subjected to herbivory actually increase their production.

Increased fitness can only be demonstrated if grazed plants are larger or form more numerous offspring at the end of the season than ungrazed plants (Westoby, 1989). (Again, we stress that not only female fitness, i.e. seed production, but also male fitness has to be examined as an important component of (over)compensation.) Reviews by Belsky (1986), drawing attention to the imperfect experimental design and statistical analyses of studies that claim overcompensation, and by Verkaar (1988), directing attention to the physiological background of overcompensation, demonstrate that beneficial effects of herbivores on individual plants are seldom found. Although numerous studies have shown a constant or increased yield in attacked plants, these studies have almost invariably focused on a single organ. Field experiments, in which removal of stem apices from *Ipomopsis aggregata* resulted in a 2.4-fold increase in seed production, did, however, focus on above-ground as well as below-ground plant parts. Comparable laboratory experiments showed that seed production was also significantly increased, though to a smaller extent. The plant seems to have evolved to depend on herbivory to achieve its greater fitness. In this special case, herbivory was restricted to only a short period at the start of the growing season, but more than 70% of plants were severely damaged (Paige and Whitham, 1987).

As pointed out by Crawley (1987), it may be unclear at first why ungrazed plants do not also have a comparable, second burst of flowering after the period of herbivory has passed. However, a simple model by van der Meijden (1990) illustrates the advantage to a genotype which does not invest all its resources in early season growth, but is able to accelerate stem growth after herbivory, compared with a genotype that invests all its resources in growth before herbivory. Herbivory simply seems to be a trigger for growth in this species (van der Meijden, 1990). Ungrazed plants do not have such a trigger, so do not respond with a second burst of flowering.

Maschinsky and Whitham (1989) found evidence for a continuum of compensatory responses to vertebrate herbivory in *Ipomopsis arizonica*. In their study, the most common response to herbivory was compensation. Overcompensation was found in grazed plants that were supplied with additional nutrients and moreover were not competing with neighbours. Undercompensation was found when nutrient availability was low and *I. arizonica* was growing in mixture with grasses. This concept of a continuum in reactions of plants to herbivory (Maschinsky and Whitham, 1989) is an important contribution to the debate concerning a possible beneficial effect of herbivores to plants.

Until now, most physiological studies on the plant's reaction to herbivory have been concerned with single mechanisms, such as increased photosynthesis or a changed allocation pattern. We challenge plant physiologists to join this discussion by extending their studies to the physiological mechanisms that may lead to overcompensation. This would give an important stimulus to the debate on overcompensation.

Acknowledgements

We wish to thank Tom de Jong, Peter Klinkhamer, Rinny Kooi, Thom van Rossum and Geert van Wirdum for critically reading the manuscript.

References

Alcock, M.B. (1964) The physiological significance of defoliation on the subsequent regrowth of grass–clover mixtures and cereals. In: *Grazing in Terrestrial and Marine Environments.* Symposium British Ecological Society. Blackwell, Oxford, pp. 25–41.

Ayres, P.G. (1991) Growth responses induced by pathogens and other stresses. In: *Responses of Plants to Multiple Stesses* (eds H.J. Mooney, E.J. Pell and W.E. Winner). Academic Press, London, pp. 227–248.

Baur Höch, B., Mächler, F. and Nösberger, J. (1990) Effect of carbohydrate demand on the remobilization of starch in stolons and roots of white clover (*Trifolium repens* L.) after defoliation. *J. Exp. Bot.* **41**, 573–578.

Belsky, A.J. (1986) Does herbivory benefit plants? A review of the evidence. *Am. Nat.* **127,** 870–892.

Benner, B.L. (1988) Effects of apex removal and nutrient supplementation on branching and seed production in *Thlaspi arvense* (Brassicaceae). *Am. J. Bot.* **75**, 645–651.

Bentley, S. and Whittaker, J.B. (1979) Effects of grazing by a chrysomelid beetle, *Gastrophysa viridula*, on competition between *Rumex obtusifolius* and *Rumex crispus. J. Ecol.* **67**, 79–90.

Bergström, R. and Danell, K. (1987) Effects of simulated winter browsing by moose on morphology and biomass of two birch species. *J. Ecol.* **75**, 533–544.

Brouwer, R. (1962a) Distribution of dry matter in the plant. *Neth. J. Agric. Sci.* **10**, 361–376.

Brouwer, R. (1962b) Influence of temperature of the root medium on the growth of seedlings of various crop plants. *Mededel. I.B.S.* **175**, 11–18.

Brouwer, R. (1963) Some aspects of the equilibrium between overground and underground plant parts. *Mededel. I.B.S.* **213**, 31–39.

Brouwer, R. (1983) Functional equilibrium: sense or nonsense? *Neth. J. Agric. Sci.* **31**, 335–348.

Caldwell, M.M., Richards, J.H., Johnson, D.A., Nowak, R.S. and Dzurec, R.J. (1981) Coping with herbivory: photosynthetic capacity and resource allocation in two semiarid *Agropyron* bunchgrasses. *Oecologia* **50**, 14–24.

Cockfield, S.D. and Potter, D.A. (1986) Interaction of *Euonymus* scale (Homoptera: Diaspididae) feeding damage and severe water stress on leaf abscission and growth of *Euonymus fortunei*. *Oecologia* **71**, 41–46.

Cottam, D.A., Whittaker, J.B. and Malloch, A.J.C. (1986) The effects of chrysomelid beetle grazing and plant competition on the growth of *Rumex obtusifolius*. *Oecologia* **70**, 452–456.

Coughenor, M.B., McNaughton, S.J. and Wallace, L.L. (1985) Responses of an African tall-grass (*Hyparrhenia filipendula* Stapf.) to defoliation and limitations of water and nitrogen. *Oecologia* **68**, 80–86.

Cox, C.S. and McEvoy, P.B. (1983) Effect of summer moisture stress on the capacity of tansy ragwort (*Senecio jacobaea*) to compensate for defoliation by the cinnabar moth (*Tyria jacobaeae*). *J. Appl. Ecol.* **20**, 225–234.

Crawley, M.J. (1987) Benevolent herbivores? *Trends Ecol. Evol.* **2**, 167–168.

Culvenor, R.A., Davidson, I.A. and Simpson, R.J. (1989) Regrowth by swards of subterranean clover. 2. Carbon exchange in shoot, root and nodule. *Ann. Bot.* **64**, 557–567.

Culvenor, R.A. and Simpson, R.P. (1991) Mobilization of nitrogen in swards of *Trifolium subterraneum* L. during regrowth after defoliation. *New Phytol.* **117**, 81–90.

Cyr, D.A. and Bewley, D.J. (1990) Seasonal variation in nitrogen storage reserves in the roots of leafy spurge (*Euphorbia esula*) and responses to decapitation and defoliation. *Physiol. Plant* **78**, 361–366.

Danell, K., Huss-Danell, K. and Bergström, R. (1985) Interactions between browsing moose and two species of birch in Sweden. *Ecology* **66**, 1867–1878.

Doak, D.F. (1991) The consequences of herbivory for dwarf fireweed: different time scales, different morphological scales. *Ecology* **72**, 1397–1407.

Dyer, M.I., Acra, M.A., Wang, G.M., Coleman, D.C., Freckman, D.W., McNaughton, S.J. and Strain, B.R. (1991) Source-sink carbon relations in two *Panicum coloratum* ecotypes in response to herbivory. *Ecology* **72**, 1472–1483.

Eaton, F.M. (1931) Early defloration as a method of increasing cotton yield, and the relation of fruitfulness to fiber and boll characters. *J. Agric. Res.* **42**, 447–462.

van Emden, H.F. (1990) The effect of *Brevicoryne brassicae* on leaf area, dry matter distribution and amino acids of the Brussels sprout plant. *Ann. Appl. Biol.* **116**, 199–204.

Fajer, E.D., Bowers, M.D. and Bazzaz, F.A. (1991) Performance and allocation patterns of the perennial herb, *Plantago lanceolata* in response to simulated herbivory and elevated CO_2 environments. *Oecologia* **87**, 37–42.

Harris, P. (1973) Insects in the population dynamics of plants. In: *Insect/plant relationships*. Symposia of the Royal Entomological Society of London, Volume 6 (ed. H.F. van Emden). Blackwell, Oxford, pp. 201–209.

Harris, P., Thomson, L.S., Wilkinson, A.T.S. and Neary, M.E. (1978) Reproductive biology of tansy ragwort, climate and biological control by the cinnabar moth. In: *Proceedings IV International Symposium on Biological Control of Weeds* (ed. I.E. Freeman), pp. 163–173.

Ingham, R.E. and Detling, J.K. (1986) Effects of defoliation and nematode consumption on growth and leaf gas exchange in *Bouteloua curtipendula*. *Oikos* **46**, 23–28.

Inouye, D.W. (1982) The consequences of herbivory: a mixed blessing for *Jurinea mollis* (Asteraceae). *Oikos* **39**, 269–272.

Islam, Z. and Crawley, M.J. (1983) Compensation and regrowth in ragwort (*Senecio jacobaea*) attacked by cinnabar moth (*Tyria jacobaeae*). *J. Ecol.* **71**, 829–843.

Jaramillo, V.J. and Detling, J.K. (1988) Grazing history, defoliation, and competition: effects on shortgrass production and nitrogen accumulation. *Ecology* **69**, 1599–1608.

de Jong, T.J. and Klinkhamer, P.G.L. (1991) Early flowering in *Cynoglossum officinale* L. constraint or adaptation? *Func. Ecol.* **5**, 750–756.

Kolodny-Hirsch, D.M., Saunders, J.A. and Harrison, F.P. (1986) Effects of simulated tobacco

hornworm (Lepidoptera: Sphingidae) defoliation on growth dynamics and physiology of tobacco as evidence of plant tolerance to leaf consumption. *Environ. Entomol.* **15**, 1137–1144.
Larsson, S. (1983) Effects of artificial defoliation on stem growth in *Salix smithiana* grown under intensive culture. *Acta Oecol. Oecol. Applic.* **4**, 343–349.
Leather, S.R. (1986) Insects on bird cherry. I. The bird cherry ermine moth, *Yponomeuta evonymellus. Entomol. Gaz.* **37**, 209–213.
Leather, S.R. (1988) Consumers and plant fitness: coevolution or competition? *Oikos* **53**, 285–288.
Louda, S.M. (1984) Herbivore effect on stature, fruiting, and leaf dynamics of a native crucifer. *Ecology* **65**, 1379–1386.
Marriot, C.A. and Haystead, A. (1990) The effect of defoliation on the nitrogen economy of white clover: regrowth and the remobilization of plant organic nitrogen. *Ann. Bot.* **66**, 465–474.
Maschinsky, J. and Whitham, T.G. (1989) The continuum of plant responses to herbivory: the influence of plant association, nutrient availability, and timing. *Am. Nat.* **134**, 1–19.
van der Meijden, E. (1990) Herbivory as a trigger for growth. *Funct. Ecol.* **4**, 597–598.
Michaud, J.P. (1989) Nectar accumulation in flowers of fireweed, *Epilobium angustifolium* (Onograceae), in response to simulated defoliation. *J. Apic. Res.* **28**, 181–186.
Neales, T.F. and Incoll, L.D. (1968) The control of leaf photosynthesis rate by the level of assimilate concentration in the leaf: a review of the hypothesis. *Bot. Rev.* **34**, 107–125.
Nooden, L.D. (1980) Senescence in the whole plant. In: *Senescence in Plants* (ed. K.V. Thimann). CRC Press, Boca Raton, FL, pp. 219–258.
Nowak, R.S. and Caldwell, M.M. (1984) A test of compensatory photosynthesis in the field: implications for herbivory tolerance. *Oecologia* **61**, 311–318.
Oesterheld, M. and McNaughton, S.J. (1991a) Effect of stress and time for recovery on the amount of compensatory growth after grazing. *Oecologia* **85**, 305–313.
Oesterheld, M. and McNaughton, S.J. (1991b) Interactive effect of flooding and grazing on the growth of Serengeti grasses. *Oecologia* **88**, 153–156.
Olson, B.E. and Richards, J.H. (1988a) Annual replacement of the tillers of *Agropyron desertorum* following grazing. *Oecologia* **76**, 1–6.
Olson, B.E. and Richards, J.H. (1988b) Spatial arrangement of tiller replacement in *Agropyron desertorum* following grazing. *Oecologia* **76**, 7–10.
Owen, D.F. and Wiegert, R.G. (1976) Do consumers maximize plant fitness? *Oikos* **27**, 488–492.
Paige, K.N. and Whitham, T.G. (1987) Overcompensation in response to mammalian herbivory: the advantage of being eaten. *Am. Nat.* **129**, 407–416.
Paul, N.D. and Ayres, P.G. (1986) The effects of infection by rust (*Puccinia lagenophorae* Cooke) on the growth of groundsel (*Senecio vulgaris* L.) cultivated under a range of nutrient concentrations. *Ann. Bot.* **58**, 321–331.
Polley, H.W. and Detling, J.K. (1989) Defoliation, nitrogen and competition: effects on plant growth and nitrogen nutrition. *Ecology* **70**, 721–727.
Prins, A.H., Verkaar, H.J. and M. van den Herik (1989) Responses of *Cynoglossum officinale* L. and *Senecio jacobaea* L. to various degrees of defoliation. *New Phytol.* **111**, 725–731.
Reichman, O.J. and Smith, S.C. (1991) Responses to simulated leaf and root herbivory by a biennial, *Tragopogon dubius. Ecology* **72**, 116–124.
Richards, J.H. (1984) Root growth response to defoliation in two *Agropyron* bunchgrasses: field observations with an improved root periscope. *Oecologia* **64**, 21–25.
Royalty, R.N. and Perring, T.M. (1989) Reduction in photosynthesis of tomato leaflets caused by tomato russet mite (Acari: Eriophyidae). *Environ. Entomol.* **18**, 256–260.

Ruess, R.W. (1988) The interaction of defoliation and nutrient uptake in *Sporobolus kentrophyllus*, a short grass species from the Serengeti Plains. *Oecologia* **77**, 550–556.

Salisbury, F.B. and Ross, C.W. (1985) *Plant Physiology*, 3rd edition. Wadsworth, Belmont, CA.

Seastedt, T.R., Crossley, D.A.Jr. and Hargrove, W.J. (1983) The effect of low-level consumption by canopy arthropods on the growth and nutrient dynamics of black locust and red maple trees in the southern Appalachians. *Ecology* **64**, 1040–1048.

Shields, E.J. and Wyman, J.A. (1984) Effects of defoliation at specific growth stages on potato yields. *J. Econ. Entomol.* **77**, 1194–1199.

Simoes, M. and Baruch, Z. (1991) Responses to simulated herbivory and water stress in two tropical C_4 grasses. *Oecologia* **88**, 173–180.

Smith, D. and Silva, J.P. (1969) Use of carbohydrate and nitrogen root reserves in the regrowth of alfalfa from greenhouse experiments under light and dark conditions. *Crop Sci.* **9**, 464–467.

Solomon, B.P. (1983) Compensatory production in *Solanum carolinense* following attack by a host-specific herbivore. *J. Ecol.* **71**, 681–690.

Stamp, N.E. (1984) Effect of defoliation by checkerspot caterpillars (*Euphydryas phaeton*) and sawfly larvae (*Macrophya nigra* and *Tenthredo grandis*) on their host plants (*Chelone* spp.). *Oecologia* **63**, 275–280.

Sweet, G.B. and Wareing, P.F. (1966) Role of plant growth in regulating photosynthesis. *Nature* **210**, 77–79.

Taylor, W.E. and Bardner, R. (1968) Effect of feeding by larvae of *Phaedon cochleariae* (F.) and *Plutella maculipennis* (Curt.) on the yield of radish and turnip plants. *Ann. Appl. Biol.* **62**, 249–252.

Thimann, K.V. (1980) The senescence of leaves. In: *Senescence in Plants* (ed. K.V. Thimann). CRC Press, Boca Raton, Florida, pp. 86–115.

Thimann, K.V. (1987) Plant senescence: a proposed integration of the constituent processes. In: *Plant Senescence: Its Biochemistry and Physiology* (eds W.W. Thomson, E.A. Nothnagel and R.C. Huffaker). Proc. X Annual Symposium in Plant Physiology, pp. 1–19.

Verkaar, H.J. (1988) Are defoliators beneficial for their host plants in terrestrial ecosystems – a review. *Acta Bot. Neerl.* **37**, 137–152.

Verkaar, H.J., van der Meijden, E. and Breebaart, L. (1986) The responses of *Cynoglossum officinale* L. and *Verbascum thapsus* to defoliation in relation to nitrogen supply. *New Phytol.* **104**, 121–129.

Vranjic, J.A. and Gullan, P.J. (1990) The effect of a sap-sucking herbivore, *Eriococcus coriaceus* (Homoptera: Eriococcidae), on seedling growth and architecture in *Eucalyptus blakelyi*. *Oikos* **59**, 157–162.

Wallace, L.L., McNaughton, S.J. and Coughenor, M.B. (1984) Compensatory photosynthetic responses of three African grassland species to different fertilization, watering, and clipping regimes. *Bot. Gaz.* **145**, 151–156.

Walters, D.R. and Ayres, P.G. (1983) Changes in nitrogen utilization and enzyme activities associated with CO_2 exchanges in healthy leaves of powdery mildew-infected barley. *Physiol. Plant Pathol.* **23**, 447–459.

Wareing, P.F., Khalifa, M.M. and Treharne, K.J. (1968) Rate limiting processes in photosynthesis at saturating light intensities. *Nature* **220**, 453–457.

Wareing, P.F. and Patrick, J. (1975) Source-sink relations and the partition of assimilates in the plant. In: *Photosynthesis and Productivity in Different Environments* (ed. J.P. Cooper). Cambridge University Press, Cambridge, pp. 481–499.

Westoby, M. (1989) Selective forces exerted by vertebrate herbivores on plants. *Trends Ecol. Evol.* **4**, 115–117.

Whittaker, J.B. (1979) Invertebrate grazing, competition and plant dynamics. In: *Population*

Dynamics (eds R.M. Anderson, B.D. Turner, L.R. Taylor). Volume 20, Symposia of the British Ecological Society. Blackwell Scientific Publications, Oxford, pp. 207–222.

Williams, P.F. (1978) Growth of broad beans infected by *Uromyces viciae-fabae*. *Ann. Appl. Biol.* **90**, 329–334.

Wit, A.K.H. (1985) The relation between partial defoliation during the preheading stages of spring cabbage and yield, as a method to assess the quantitative damage induced by leaf-eating insects. *Z. Ang. Entomol.* **100**, 96–100.

Woolhouse, H.W. (1967) The nature of senescence in plants. In: *Aspects of the Biology of Aging*, Volume 21, Symposium for the Society for Experimental Biology (ed. H.W. Woolhouse). Cambridge University Press, Cambridge, pp. 179–213.

3

Signals in the wounded plant

Dianna Bowles

3.1 Introduction

Plants have evolved a range of defence mechanisms for protection against pests and pathogens. Some of these are constitutive and the mechanism is an in-built feature of vegetative growth and development of the plant. For example, trichomes (glandular cells and cell communities of the epidermis) represent a structural constitutive defence mechanism protecting the aerial regions of most plant species. Other defence mechanisms are inducible, such that challenge by a pest or pathogen triggers a response that leads to the establishment of a defence state. These mechanisms may involve gene products already *in situ*, for example the local induction of callose synthase activity at a site of challenge (Kauss, 1990), or the induction of glucanases and chitinases or proteinase inhibitors. Alternatively, inducible mechanisms may lead to the activation of defence genes. Defence genes may encode products that function directly, for example as deterrents or anti-feedants, or the genes may encode enzymes that synthesize defence–related products such as phytoalexins (Bowles, 1990). Many of the changes regulated during a defence response are also regulated developmentally. Often, the same gene, as well as different members of a gene family, is subject to both developmental regulation and environmental regulation: developmental, when a particular stage in the life cycle of a plant requires protection, such as a storage organ or a seed; environmental, when a stimulus is perceived and a defence state is required. One stimulus, such as a challenge by a pathogen, pest, or predator, leads to a broad spectrum of changes that must be co-ordinated for maximum efficiency within the plant. It is now fully accepted that the local stimulus triggers signalling events that affect the whole plant, not merely the area locally affected. Therefore, the co-ordination of these local and systemic events must arise through extensive networking of signals and signal transduction pathways (Bowles, 1990).

Interest in Leeds concerns the networking involved in the co-ordination of a wound response (Bowles, 1991) and the response of tomato plants to physical injury has been used as the experimental system. Although we are not using a foliar pest or pathogen to trigger the response, the interactions of these organisms with the plant undoubtedly lead to 'damage', which may be perceived as a stimulus that is nearly identical to that produced aseptically by injury. For example, we know that the 'damage' produced in the root system by a nematode migrating through the cortex on its way to set up a feeding site is sufficient to trigger a systemic wound response that can be assayed in the leaves (Bowles *et al.*, 1991). In this interaction, the damage is minimal and involves only a single file of cells within the root cortex through which the nematode cuts its way as it migrates towards the vascular tissue to establish a feeding site. Despite the lack of any considerable injury, the local stimulus is sufficient to lead to increases in steady-state levels of defence gene transcripts throughout the leaves within 6 h of challenge. Therefore, we consider that conclusions drawn from data using this model system and aseptic injury will be applicable to responses induced by damage caused by pests and pathogens. The following discussion concerns data arising primarily from our work at Leeds. For wider reviews, the reader is referred to Kahl (1982), Bol *et al.* (1990), Chessin and Zipf (1990), and Ryan and Farmer (1991).

3.2 Gene markers for study of the wound response

We have used two gene markers to follow local and systemic wound-induced events in tomato plants. The cDNA, *pTOM13*, corresponds to the gene that encodes ethylene-forming enzyme (Hamilton *et al.*, 1991). The cDNA, *pin 2*, corresponds to the gene encoding the serine proteinase inhibitor 2 (Green and Ryan, 1972).

We initially chose to work with *pTOM13* because of its known induction by wounding: steady-state levels of transcripts were shown to increase rapidly in wounded leaf tissue within 10 min of shredding the leaf and assaying the segments (Smith *et al.*, 1986; Holdsworth *et al.*, 1987). At the time we began our studies, Grierson and co-workers at Nottingham were studying these events, but the function of the gene product corresponding to *pTOM13* was not known. Recently, the same group has demonstrated, using a number of approaches, that the function is related to the formation of ethylene and, specifically, *pTOM13* most probably corresponds to the ethylene-forming enzyme (ETE, see equation) (Hamilton *et al.*, 1990, 1991; Spann *et al.*, 1991). Therefore the use of *pTOM13* as a gene marker provides an insight into the spatial and temporal regulation of ethylene biosynthesis during a wound response, particularly if the use of that marker is coupled with a cDNA encoding 1-aminocyclopropane-1-carboxylic acid (ACC) synthase (Van der Straeten *et al.*, 1990).

$$\begin{matrix}\text{S-adenosyl} & \xrightarrow{\text{ACC synthase}} & \text{1-aminocyclopropane-} & \xrightarrow{\text{ETE}} & \text{ethylene} \\ \text{methionine} & & \text{1-carboxylic} & & \\ \text{(SAM)} & & \text{acid (ACC)} & & \end{matrix}$$

Proteinase inhibitors are known anti-feedants and have been shown to exhibit toxic effects when assayed directly as protein additives in artificial diets or when the genes encoding the inhibitors have been expressed in transgenic plants (Bowles, 1990). The serine proteinase inhibitors 1 and 2, such as those studied widely as markers for the systemic wound response of potato and tomato plants, are expressed only in the aerial regions of the vegetative plant and presumably therefore have evolved as an inducible defence mechanism against foliar pests and pathogens.

3.3 Co-ordination of wound-induced events

Despite a number of studies of *pin* induction during a wound response, very few attempts have been made to relate the pattern of *pin* induction with that of other wound-induced genes, such as those encoding enzymes of ethylene biosynthesis. Our experimental system uses tomato seedlings bearing cotyledon leaves and two fully expanded leaves. Mechanical injury to the terminal leaflet of leaf 1 on the plant can lead both to local events, that is, in an area within 0.5 cm of the wound, and to systemic events, it passes to (1) the unwounded regions of the terminal leaflet of leaf 1, (2) the subterminal leaflets of leaf 1, and (3) leaf 2 and other distant tissues of the plant. It may well be possible that the signalling mechanism(s) across the lamina of a single leaflet is different from that involved in leaflet-to-leaflet signalling, and is different again to that involved in setting up a systemic response in a second leaf and elsewhere in the plant.

In this respect, our findings on the wound-induced expression patterns of *pTOM13* and *pin 2* may be relevant (Doherty *et al.*, manuscript submitted). Consider first the expression of *pin 2*, as detected by changes in steady-state levels of mRNA in Northern analyses. A wound to the terminal leaflet invariably leads to increased levels of transcripts at the wound-site and also, within 4 h, these are detected in systemically responding unwounded cells of the same leaflet. A systemic response in the subterminal leaflet is also generally observed by 4 h, but the range of the systemic response would seem to be dependent on the area of local damage. Provided the surface area is large, relative to the size of the terminal leaflet, transcript levels increase in leaf 2, if not, the subterminal leaflet responds systemically but leaf 2 does not. When the cell-specific expression of *pin 2* is analysed with *in situ* hybridization, *pin 2* transcripts can be detected throughout the internal cells of the lamina but not in the epidermis. This pattern is identical within the wounded leaflet, the unwounded leaflet, and the more distant leaf 2.

The temporal and spatial pattern of wound-induced expression of *pin 2* contrasts markedly with the expression of *pTOM13*. In Northern analyses, steady-state levels of mRNA to *pTOM13* are found to increase only in the local injury zone, not elsewhere in the wounded leaflet, nor elsewhere in the plant. Thus, from this information alone, *pTOM13* could be thought of as an excellent marker for a local wound response, and a rapid response, since transcript levels increase within 10 min of injury, peak at 1 h and decrease to neglible levels by 2 h. However, when the cell-specific expression of *pTOM13* is analysed with *in situ* hybridization, *pTOM13* transcripts can be detected outside the injury zone. Specifically, *pTOM13* transcripts become detectable in each of the veins within the wounded leaflet. No signal can be detected in unwounded leaflets, but nevertheless the expression of *pTOM13* can be defined as local and systemic, since transcript levels do increase in discrete cell populations of the leaflet beyond the damage site. Thus, a wound stimulus perceived at a local region of the terminal leaflet leads to signalling event(s) that differentially regulate the temporal and spatial expression patterns of the two genes.

3.4 Differential effects of glycan elicitors

In this context of gene markers for wound-induced events, we have shown a differential effect of cell wall fragment elicitors on the expression of the same two genes. Application of pectic fragments (degree of polymerization (dp) 10–20) to excised tomato plants routinely leads to increased steady-state levels of *pin 2* transcripts. This elicitor mixture, however, does not induce *pTOM13* expression. In contrast, application of small oligogalacturonides (degree of polymerization (dp) 2–9), and of a purified trimer, leads to increased steady-state levels of *pTOM13* transcripts but has no effect on *pin 2* expression (Doherty *et al.*, manuscript submitted). Is it possible, therefore, that during a wound response *in planta*, physical injury to the leaf blade leads to hydrolysis of cell wall polysaccharides at the injury zone and the resultant fragments have diverse effects, dependent in their size and structures? In this model, the fragments do not represent the systemic signal, but rather represent a range of local signals capable of setting up different events that each lead to different systemic consequences.

3.5 Rapid events induced by glycan elicitors

In earlier collaborative work (Thain *et al.*, 1990), both pectic fragments (dp 10–20) and the smaller oligogalacturonides (dp 2–9) were shown to induce rapid membrane depolarization. The pattern of depolarization was different for the two fragments since, in contrast to the oligogalacturonides, the response to pectic fragments was biphasic, with an immediate change in membrane potential being followed by a slower change (Figure 3.1). These studies were carried out using intracellular electrodes and an assay system involving tomato leaf mesophyll cells in a leaf segment stripped of epidermis. In a subsequent study employing different techniques and using tobacco suspension cultured cells, Guern and coworkers have extended the approach (Mathieu *et al.*, 1991). In their experimental system, small oligogalacturonides of dp < 6 had no effect on membrane potential, whereas the larger fragments caused depolarization. The depolarization was accompanied by potassium efflux, a decrease in cytoplasmic pH and a decrease in external calcium.

In order to understand the implications of these membrane-associated events induced by cell wall fragments, it will be necessary to identify the membrane components recognizing the different signals. The most efficient means of achieving this may well be the use of anti-idiotype antibodies, an approach similar to that used in the identification of potential receptors to classical growth regulators.

3.6 Local events and systemic signalling

The changes in ion fluxes induced by cell wall fragments may reflect local intracellular signalling events, linking recognition of the oligosaccharide at the cell surface to changes in gene activation. Alternatively (and/or additionally), the ion fluxes may be involved in the setting up and establishment of a systemic signalling system. In this respect, electrical signals are known to be generated by a heat stimulus (Wildon *et al.*,

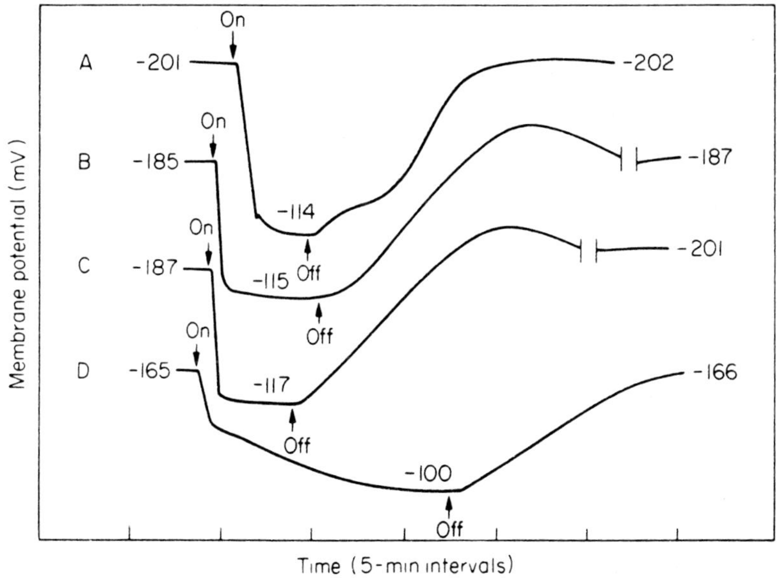

Figure 3.1*. Depolarization of membrane potential of tomato leaf mesophyll cells by NaCN (1 mol m^{-3}) and by oligosaccharides (1 g dm^{-3}), all in standard medium at pH 5.5. Arrows indicate times of addition (on) and removal (off) of NaCN and oligosaccharides. Numbers indicate the values of the membrane potential in mV just before treatment, at the maximum depolarization, and after recovery. Traces A, B and C are parts of a continuous measurement on the same cell in the time sequence B, C, A. A shows the depolarization caused by NaCN. B and C are the depolarizations caused by two successive treatments with oligogalacturonides. Trace D shows a depolarization caused by pectic fragments. In traces B and C, a 10 min portion during which the membrane potential approached the new steady value (−187 mV and −201 mV) has been omitted. The transient hyperpolarizations shown in traces B and C during recovery from oligogalacturonides were not a common feature. Reproduced from Thain* et al. *(1990) with permission from Blackwell Scientific.*

1989, and references therein). More recently, again in collaboration with Thain and Wildon, we have shown that physical injury to a cotyledon leaf of a tomato plant leads to propagated action potentials that can be detected by surface electrodes on the petiole and stem of the injured plant (Wildon *et al.*, 1992). A systemic response of this type is well characterized for many lower animals, where the spread is propagated via electrical coupling of the cells. A similar rapid response is shown in the specialized epithelial sheets of higher animals, where the individual cells are coupled through gap junctions. It is possible, therefore, that injury sets up action potentials that are propagated through plant tissue, or that injury leads to the release of cell wall fragments and the effects of these on ion transport at the local site may in turn lead to systemic electrical signals.

The means by which the local signals arising from tissue damage may lead to systemic changes in gene expression will be an area of intense research interest for many years to come. For a full discussion of the many models proposed as the signalling mechanisms, the reader is referred to Bowles (1991).

Acknowledgement

This work was supported by the AFRC.

References

Bol, J.F., Lindhorst, H.J.M. and Cornelissen, B.J.C. (1990) Plant pathogenesis-related proteins induced by virus infection. *Ann. Rev. Phytopathol.* **28**, 113–138.

Bowles, D.J. (1990) Defence-related proteins in higher plants. *Ann. Rev. Biochem.* **59**, 873–907.

Bowles, D.J. (1991) The wound response of plants. *Curr. Biol.* **1**, 165–167.

Bowles, D.J., Hammond-Kosack, K., Gurr, S.J. and Atkinson, H.J. (1991) Local and systemic changes in plant gene expression following root infection with cyst nematodes. In: *Biochemistry and Molecular Biology of Plant-Pathogen Interactions* (ed. C.J. Smith). Oxford University Press, pp. 225–236.

Chessin, M. and Zipf, A.E. (1990) Alarm systems in higher plants. *Bot. Rev.* **56**, 193–235.

Green, T.R. and Ryan, C.A. (1972) Wound-induced proteinase inhibitor in plant leaves: a possible defense mechanism against insects. *Science* **175**, 776–777.

Hamilton, A.J., Bonzayer, M. and Grierson, D. (1991) Identification of a tomato gene for the ethylene-forming enzyme by expression in yeast. *Proc. Natl. Acad. Sci. USA* **88**, 7434–7437.

Hamilton, A.J., Lycett, G.W. and Grierson, D. (1990) Antisense gene that inhibits synthesis of the hormone ethylene in transgenic plants. *Nature* **346**, 284–287.

Holdsworth, M.J., Bird, C.R., Schuch, R.J. and Grierson, W.D (1987) Structure and expression of an ethylene-related mRNA from tomato. *Nucl. Acids Res.* **15**, 731–739.

Kahl, G. (1982) Molecular biology of wound healing. In: *Molecular Biology of Plant Tumors*. Academic Press, New York, pp. 211–267.

Kaus, H. (1990) Role of the plasmamembrane in host-pathogen interactions. In: The Plant Plasmamembrane (eds C. Larsson and I.M. Moller). Springer-Verlag, Berlin, pp. 321–350.

Mathieu, Y., Kurkdjian, A., Xia, H., Guern, J., Koller, A., Spiro, M.D., O'Neill, M., Albersheim, P. and Darvill, A. (1991) Membrane responses induced by oligogalacturonides in suspension-cultured tobacco cells. *Plant J.* **1**, 333–343.

Ryan, C.A. and Farmer, E.E. (1991) Oligosaccharide signals in plants. *Ann. Rev. Plant Physiol. Plant Mol. Biol.* **42**, 651–674.

Smith, C.J.S., Slater, A. and Grierson, D. (1986) Rapid appearance of an mRNA correlated with ethylene synthesis encoding a protein of molecular weight 35 000. *Planta* **168**, 94–100.

Spann, P., Reinhart, D. and Boller, T. (1991) Analysis and cloning of the ethylene-forming enzyme from tomato by functional expression of its mRNA in *Xenopus laevis* oocytes. *EMBO J.* **10**, 2007–2013.

Thain, J., Doherty, H., Bowles, D.J. and Wildon, D. (1990) Oligosaccharides that induce proteinase inhibitors in tomato plants cause depolarization of tomato leaf cells. *Plant Cell Environ.* **13**, 569–574.

Van der Straeten, D., Van Wiemeersch, L., Goodman, H.M. and Van Montagu, M. (1990) Cloning and sequence of two different cDNAs encoding 1-aminocyclopropane-1-carboxylate synthase in tomato. *Proc. Natl. Acad. Sci. USA* **87**, 4859–4863.

Wildon, D.C., Doherty, H., Eagles, G., Bowles, D.J. and Thain, J.F. (1989) Systemic proteinase inhibitor induction and propagated electrical responses arising from localized heat stimuli in tomato plants. *Ann. Bot.* **646**, 691–695.

Wildon, D.C., Thain, J.F., Minckin, P.E.H., Gubb, I., Reilley, A., Skipper, Y., Doherty, H., O'Donnell, P. and Bowles, D.J. (1992) Electrical signalling and systemic proteinase inhibitor synthesis in the wounded plant. *Nature* in press.

4

Resistance: the role of the hypersensitive response and phytoalexins

R.N. Strange

4.1 Definitions and introduction

The hypersensitive response (HR) is the rapid but localized death of plant cells in response to challenge by avirulent micro-organisms. This reaction is associated with both local and systemic acquired resistance of the plant which may be mediated through the deposition of callose, lignin and hydroxyproline-rich glycoproteins as well as the synthesis of chitinases, glucanases and phytoalexins. Salicylic acid is a candidate for the signal compound that initiates these multiple reactions (Malamy *et al.*, 1990; Métraux *et al.*, 1990; Ward *et al.*, 1991), although recently published evidence suggests that salicylic acid itself accumulates in response to an earlier but unknown signal (Rasmussen *et al.*, 1991).

Phytoalexins are low molecular weight anti-microbial compounds that are both synthesized by and accumulate in plants in response to infection (Paxton, 1981). They are normally synthesized in plant cells adjacent to cells that are dying or dead, such as those responding hypersensitively, and accumulate in the moribund cells (Bailey, 1982). There is now considerable evidence that supports their role as an important component of resistance to pathogenic organisms in a number of plants, some of which will be reviewed in this chapter.

Factors that initiate the HR, as well as the phytoalexin and other defence responses, are referred to as elicitors. Abiotic elicitors may be physical, such as UV light or wounding, or chemical, such as the salts of heavy metals. In contrast, biotic elicitors comprise a wide range of compounds that originate from living organisms.

4.2 The hypersensitive response

4.2.1 *Genetic perspective*

Two distinct genetic aspects of the HR are currently recognized. One of these is that the HR results from the interaction of specific plant genes for resistance (*R* genes) with specific parasite genes for avirulence (*avr* genes) in a gene-for-gene relationship (Flor, 1971; Keen, 1990; Jenner *et al.*, 1991). The other is that parasite genes eliciting the HR are pleiotropic in that they are required for pathogenicity in their homologous hosts. These genes are known as *hrp* genes since they are responsible for both the *h*ypersensitive *r*esponse and *p*athogenicity (Lindgren *et al.*, 1986, 1988). Transformation of bacterial pathogens is technically easier than that of fungi and viruses, so most of the advances in our understanding of *avr* and *hrp* genes comes from work with bacteria.

Avirulence genes. In pioneering experiments, Staskawicz *et al.* (1984) cloned an avirulence gene from *Pseudomonas syringae* pv. *glycinea* race 6 and mobilized it into three other races. The transformants acquired the avirulence characteristics of race 6 and these were superimposed on their own avirulence spectra when tested on seven soybean cultivars (Table 4.1).

Staskawicz and co-workers have also studied bacterial spot of pepper caused by *Xanthomonas campestris* pv. *vesicatoria* (Kearney *et al.*, 1988). They found 13 spontaneous mutants which had overcome the HR resistance of the host conferred by the resistance gene *Bs1*, suggesting that the corresponding avirulence gene *avrBs1* was no longer effective. Since they had previously cloned *avrBs1*, they were able to transform the virulent mutants with an effective copy of the gene. Plants challenged with the

Table 4.1. *Reactions of seven soybean cultivars to three races of* Pseudomonas syringae *pv.* glycinea *and the alteration of these reactions by transformations of the pathogen with a plasmid containing an* avr *gene from race 6 (adapted from Staskawicz* et al.*, 1984)*

	Soybean cultivars						
	1	2	3	4	5	6	7
Race 5 wild-type	–	–	–	+	+	–	+
Race 5 wild-type + plasmid	–	–	–	+	–*	–	–*
Race 6 wild-type	–	–	+	+	–	+	–
Race 4 wild-type	+	+	+	+	+	+	+
Race 4 wild-type + plasmid	–*	–*	+	+	–*	+	–*
Race 6 wild-type	–	–	+	+	–	+	–
Race 1 wild-type	+	–	+	–	–	–	+
Race 1 wild-type + plasmid	–*	–	+	–	–	–	–*
Race 6 wild-type	–	–	+	+	–	+	–

– Denotes an incompatible reaction; + denotes a compatible reaction. *Denotes change to avirulence characteristic of race 6.

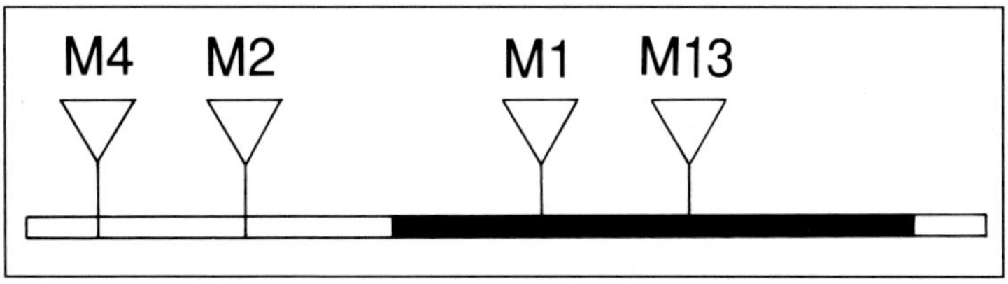

Figure 4.1. *Avirulence gene* avrBs1 *of* Xanthomonas campestris *pv.* vesicatoria *showing the position of an insertion sequence giving altered reactions in the host plant, pepper. Interaction of isolates possessing the wild-type gene with pepper cultivars containing the corresponding resistance gene* Bs1 *leads to the hypersensitive response (HR). M4, M2, M1 and M13 represent the positions of the insertion sequence IS476 in four mutants. In mutants M1 and M13 the insertions are in the open reading frame (shaded) of the* avrBs1 *gene. In these the HR was completely disrupted and the mutants caused the water-soaked lesions typical of virulent isolates of the pathogen. In M4 and M2, IS476 was found upstream from the* avrBs1 *coding region. Inoculation of plants containing* Bs1 *with M4 resulted in a delayed HR and inoculation with M2 gave rise to partially necrotic and partially water-soaked lesions (modified from Kearney* et al., *1988).*

transconjugants reacted with a typical hypersensitive response, showing that the avirulence gene had been transferred and was active in the formerly virulent mutants.

Four of the mutants with the defective *avrBs1* gene were singled out for intensive study. Analysis of the defective genes in each mutant showed that they contained an insertion sequence (IS476) of 1225 bp. In two of the mutants the insertion sequence was found upstream to the *avrBs1* gene. In one of these the HR was delayed, while in the other a reaction intermediate between the HR and the susceptible water-soaked response was obtained, when they were inoculated on pepper plants with the *Bs1* gene. The authors suggested that insertion in this upstream region affected the regulation of the gene and may have resulted in lower protein levels. In the other two mutants, which were virulent, the insertion sequence was found in the structural gene itself. So far, all virulent mutants of *avrBs1* have been found to contain the insertion sequence in this gene (Figure 4.1).

Genetic studies have linked the occurrence of IS476 with copper resistance. Copper sprays are the normal method used to control bacterial spot of pepper in Florida and almost all local strains of the bacterium are copper resistant and carry IS476. Both the gene for copper resistance and *avrBs1* are found on a 200 kb plasmid (pXvCu1) which also carries three copies of IS476, at least one of which is an actively transposable element. Thus, a measure used to control a disease, copper sprays, has resulted in the propagation of a highly mutagenic element which enables the pathogen to overcome the genetic resistance of its host (Kearney *et al.*, 1988).

In a further exciting development of the work with *X. campestris* pv. *vesicatoria*, Kearney and Staskawicz (1990) have studied another avirulence gene, *avrBs2*. The gene was found in all strains of the tomato and pepper races commonly used by the authors, as well as in 500 other strains. When *avrBs2* was mutated by the insertion of a transposon, the mutants grew only to low titres in a susceptible host (i.e. one lacking the corresponding resistance gene *Bs2*), suggesting that possession of a functional *avrBs2* gene was necessary for the fitness of the bacterium. Significantly, *Bs2* has been effective in conferring resistance of pepper to bacterial spot for many years. Thus it appears that durable resistance may result if resistance genes interact with avirulence genes which have the pleiotropic effect of contributing significantly to the fitness of the pathogen.

Organization and structure of avirulence genes. Sequence data are available for at least seven bacterial *avr* genes. These are *avrA* (Napoli and Staskawicz, 1987), *avrB* and *avrC* (Tamaki *et al.*, 1988) from *Pseudomonas syringae* pv. *glycinea*, *avrD* from *Pseudomonas syringae* pv. *tomato* (Kobayashi *et al.*, 1989), *avrBs1* (Ronald and Staskawicz, 1988) and *avrBs3* (Bonas *et al.*, 1989) from *Xanthomonas campestris* pv. *vesicatoria*, and *avrPph3* from *Pseudomonas syringae* pv. *phaseolicola* (Jenner *et al.*, 1991). An avirulence gene from *Ps. syringae* pv. *tomato* was found to be identical to *avrA* (Kobayashi *et al.*, 1989). Similarly, sequence data for a gene that determines avirulence of *X. campestris* pv. *vesicatoria* for tomato showed the gene to be almost identical to *avrBs3* (Canteros *et al.*, 1991). However, Jenner *et al.* (1991) found no homology between *avrPph3* and *avrA*, *avrB*, *avrC*, *avrD*, *avrBs1* or *avrBs3*. Recently, *Xanthomonas campestris* pv. *malvacearum*, strain H was found to contain at least six avirulence genes clustered on a 90 kb plasmid (De Feyter and Gabriel, 1991).

Ronald and Staskawicz (1988) showed that the *avrBs1* gene encoded a 50 kDa protein which had extensive homology with the carboxyl end of the 100 kDa protein coded by *avrA*. Similarly, Tamaki *et al.* (1988) found that the protein products of *avrB* and *avrC* were 42% homologous.

The discovery of homology among the several avirulence genes so far studied suggests that they are conserved. This in turn suggests that they played, and in the instance of *avrBs2* continue to play, a significant role in the fitness of the organisms that possess them. It would be interesting to know the function of such genes and how organisms in which they are apparently gratuitous are able to dispense with them.

Relation of* avr *genes to the HR. One expectation of avirulence gene products was that they would act as elicitors of defence reactions in plants in a gene-for-gene manner. This expectation has remained largely unfulfilled with bacterial pathogens as, with the possible exception of *avrBs3,* they code for cytoplasmic proteins. Moreover, no HR was elicited when the protein products of *avrC* and *avrD* were injected into soybean plants with the corresponding resistance genes (Tamaki *et al.*, 1988; Keen *et al.*, 1990). However, when *avrD* was introduced into *Escherichia coli* or *P. syringae* pv. *glycinea,* it caused the recipients to produce a low molecular weight compound that acted as a race-specific elicitor (Keen *et al.,* 1990).

In one fungal and one viral pathogen/host combination, genes that code for avirulence have been shown to be responsible for the production of elicitors. Tomatoes with the Cf9 genotype react hypersensitively to races of the fungal pathogen, *Cladosporium fulvum,* containing the avirulence gene *avr9*. The race-specific elicitor was isolated and identified as a heat-stable peptide of 27 amino acids (Schottens-Toma and DeWit, 1988). Recently de Wit and co-workers have cloned *avr9* (Van Kan *et al.,* 1991).

Oligonucleotide probes were designed according to the amino acid sequence of the elicitor and used to detect mRNA corresponding to *avr9*. Significant hybridization occurred only to mRNA prepared from the fungus containing the gene and grown *in planta*. There was no significant hybridization to mRNA preparations from such races grown *in vitro* and none to mRNA from races that did not contain *avr9*. The probe was also used to screen a cDNA library of a compatible interaction in which *avr9* was expressed. One clone was obtained that encoded the *avr9* elicitor but the single copy gene was absent from all races of the pathogen that were virulent to Cf9 tomato genotypes.

The coat protein of tobacco mosaic virus has recently been shown unequivocally to determine the HR in tobacco. Although mutants with single amino acid substitutions in the coat protein strongly indicated that this component of the virus was responsible, it was difficult to eliminate the possibility that some other property of the virus was involved, such as the components of replication. In order to overcome this problem, Culver and Dawson (1991) transformed tobacco plants with open reading frames of coat proteins that were either elicitors or non-elicitors. Transgenic plants expressing the non-elicitor coat protein did not differ from control plants but those expressing elicitor coat protein had reduced growth, were chlorotic and developed necrotic patches that coalesced resulting in the collapse of whole leaves. Moreover, the necrotic response in plants expressing a strong elicitor coat protein was more severe than in those expressing a weak elicitor coat protein.

hrp *genes.* Initial studies by Lindgren *et al.* (1986, 1988) with *Ps. syringae* pv. *phaseolicola* showed that mutation by transposon insertion of genes required for the HR also caused attenuation of the mutant on its homologous host. Since then, *hrp* genes have been described in a wide range of *Ps. syringae* pathovars as well as in *Ps. solanacearum*, *Erwinia amylovora* (Bauer and Beer, 1991) and pathovars of *X. campestris* (reviewed by Willis *et al.*, 1991).

hrp genes are normally found in clusters. For example, the *Ps. syringae* pv. *phaseolicola* cluster spans a region of 22 kb and contains nine complementation groups (Rahme *et al.*, 1991). Similarly, the *hrp* genes of *Ps. syringae* pv. *syringae* strain 61, defined by Tn*5* mutagenesis, could be complemented by a cosmid clone, pHIR11, containing a 36 kb insert (Huang *et al.*, 1988). Recent work by Huang *et al.* (1991) showed that the cosmid contained 13 complementation groups and enabled the non-pathogen, *Ps. fluorescens*, to elicit the HR in tobacco leaves. Two of these complementation groups produced either transmembrane or exported proteins (Huang *et al.*, 1991).

Work with *X. campestris* showed that all pathovars of this organism that were tested contained DNA which hybridizes with an *hrp* cluster in *Ps. solanacearum* (Arlat *et al.*, 1991). In common with other *hrp* genes, their expression, which was studied by *lacZ* fusions, was repressed by rich media. Another important finding of this study was that transcription of *hrp* genes was not regulated by genes that control the synthesis of extracellular enzymes which are also required for pathogenicity, Hrp^- mutants producing wild-type levels of these enzymes. Thus, there appear to be two independent sets of genes required for pathogenicity, *hrp* genes and genes that regulate the production or export of extracellular enzymes.

Resistance genes. We still know little about the organization and expression of resistance genes in plants. The classical genetic studies of Flor (1971) clearly showed that they interact gene-for-gene with the corresponding avirulence genes in parasites and these studies have received abundant confirmation from the molecular work on avirulence genes, some of which is summarized above. We are also ignorant about the means by which plants interact initially with non-homologous parasites and the genes responsible, although the HR is a clear sequel to this interaction. The phenomenon of *hrp* genes demonstrates that, whatever the mechanism, the plant detects a factor(s) which is necessary for pathogenicity on the homologous host. As Keen (1990) points out, this is an effective mechanism for pathogen surveillance.

4.2.2 *Elicitors*

Abiotic elicitors. Many physical and chemical insults are effective elicitors. Examples include the salts of heavy metals, UV irradiation, partial freezing, DNA intercalating compounds and free radicals. All cause membrane perturbation and gene derepression although, at least in earlier work, the parameter commonly measured was phytoalexin accumulation. For example, Epperlein *et al.* (1986) studied the elicitation of phytoalexins in cotyledons of soybean, groundnut and pea by silver nitrate, which can cause the release of free radicals. They found that the response was suppressed by scavengers of the hydroxyl radical ($OH^{\cdot}$) and suggested that the hydroxyl radical caused a chain reaction of lipid peroxidation according to the following scheme:

$$OH^{\cdot} + RH \rightarrow R^{\cdot} + H_2O$$

$$R^{\cdot} + O_2 \rightarrow ROO^{\cdot}$$

$$ROO^{\cdot} + R_1H \rightarrow ROOH + R_1^{\cdot},$$

where R and R_1 are fatty acids in membranes and $R^{\cdot}$ is a fatty acid radical.

Such a chain reaction could give rise to considerable membrane damage as occurs in the hypersensitive response and might allow the diffusion of a second messenger (see Section 4.2.3).

Biotic elicitors. Many elicitors have been described from a variety of living materials. They include polypeptides, enzymes, polysaccharides, glycoproteins, and fatty acids, as well as some low molecular weight organic compounds.

The first elicitor to be characterized was monilicolin A, a polypeptide from the fungus *Monilinia fructicola* which causes a soft rot of stone fruit. It specifically induced the phytoalexin phaseollin (Figure 4.2) in the non-host bean at concentrations as low as 10^{-9} M (Cruickshank and Perrin, 1968). The race-specific elicitor of *C. fulvum* (see Section 4.2.1) is also a polypeptide. Other proteins that have elicitor activity are enzymic. For example, a protein from *Rhizopus stolonifer* that elicited casbene (Figure 4.2), the phytoalexin of castor bean, *Ricinus communis*, co-purified to homogeneity with polygalacturonase activity (Lee and West, 1981a,b). Both elicitor and enzyme activity were destroyed by heat treatment. However, if the elicitor/polygalacturonase were allowed to interact with plant cell walls, a heat-stable elicitor was produced. The active product was identified as a mixture of oligogalacturonides. Subsequent work showed that a chain length of 13 galacturonide residues was optimal for elicitor activity while chains with less than nine residues were inactive (Amin *et al.*, 1986; Davis *et al.*, 1986).

Various saccharide preparations of the cell walls from fungi such as the soybean pathogen, *Phytophthora megasperma* f.sp. *sojae*, are elicitors. A recent study (Cheong *et al.*, 1991) has determined the structure–activity relationships of such saccharides and established that the following motif was necessary for an oligo-β-glucoside to have high elicitor activity:

$$\begin{array}{l}\mathrm{Glc}\beta(1\rightarrow6)\mathrm{Glc}\beta(1\rightarrow6)\mathrm{Glc}\beta(1\rightarrow6)\mathrm{Glc}\beta(1\rightarrow6)\mathrm{R.}\\ \mathrm{Glc}\beta(1\rightarrow3)\!\lrcorner \qquad\qquad \mathrm{Glc}\beta(1\rightarrow3)\!\lrcorner\end{array}$$

By the use of a radioactively labelled tyramine residue as the R group of the above structure, Cheong and Hahn (1991) were able to show that the microsomal membranes of soybean have a specific, high-affinity binding site for the molecule. In another study, Barber *et al.* (1989) showed that chitin oligomers with a degree of polymerization $\geqslant 3$ were active elicitors of hypersensitive lignification in wheat.

Kogel *et al.* (1988) have also studied the hypersensitive lignification of wheat. They isolated a glycoprotein from *Puccinia graminis* f. sp. *tritici* with a molecular weight of 67 kDa and a carbohydrate portion consisting predominantly of mannose and galactose. The carbohydrate portion of the molecule was deemed necessary for activity since digestion of the protein moiety by trypsin or pronase had no effect. Recent studies of this system have also shown that the elicitor binds to the plasma membrane of the plant. Binding was not, however, related to specificity of the pathogen for its host since binding occurred with membranes of wheat, irrespective of the presence of the *Sr5* allele for resistance, and also with plasma membranes of barley (Kogel *et al.*, 1991).

Bostock *et al.* (1981) found that a lipid fraction from the cystospores of *Phytophthora infestans* was an effective elicitor of the terpenoid phytoalexins of potato (Figure 4.2). They identified eicosapentaenoic and arachidonic acids as the active compounds and found that their activity was enhanced by fungal glucans (Maniara *et al.*, 1984). Both acids would be expected to be subject to peroxidation and perhaps this chain reaction is transferred to lipids making up the plasma membrane of the host, thus causing the HR.

Since diverse compounds and stimuli give rise to defence reactions in plants, it seems probable that there is some type of second messenger which signals the initiation of gene transcription and translation leading to these responses. Various candidates have been suggested according to the plants used in such experiments. For example, when water droplets containing sucrose in the μg cm^{-3} range were added to superficially wounded leaves of pigeonpea, three phytoalexins, pinostrobin chalcone and two stilbene carboxylic acids (Figure 4.2), were synthesized (Cooksey *et al.*, 1983). More recently, the expression of the *CHS-A* gene for chalcone synthase from *Petunia* was found to be sugar dependent, with sucrose being the most effective of six sugars tested (Tsukaya *et al.*, 1991). Several studies have shown that glutathione elicits the phytoalexin response when introduced to suspension cell cultures of *Phaseolus vulgaris*

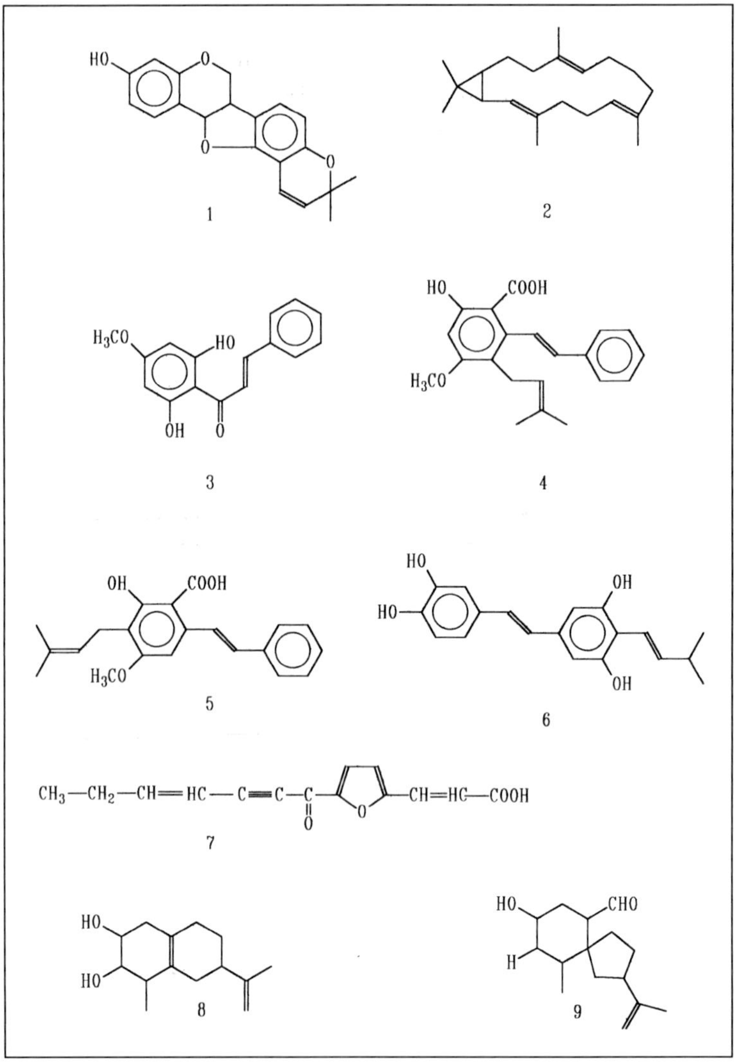

Figure 4.2*. Some representative structures of phytoalexins that are mentioned in the text. 1, phaseollin from French bean; 2, casbene from castor bean; 3, pinostrobin chalcone from pigeonpea leaves; 4 and 5, two isomeric stilbene-2-carboxylic acids from pigeonpea leaves; 6, arachidin I from groundnut cotyledons; 7, wyerone acid from broad bean; 8 and 9, rishitin and lubimin, respectively, from potato; 10, 11 and 12, the avenalumins, three nitrogen-containing phytoalexins from oats; 13 and 14, methoxybrassinin and cyclobrassinin, two sulphur-containing phytoalexins from oilseed rape; 15, 6-methoxymellein from carrot; 16, resveratrol, found in groundnut cotyledons and grapevine leaves; 17, pterostilbene from grapevine leaves; 18, glyceollin I from soybean; 19, momilactone A from rice.*

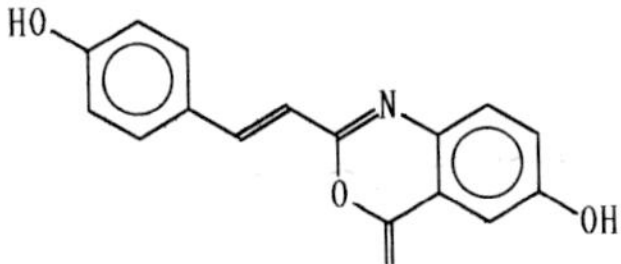

10

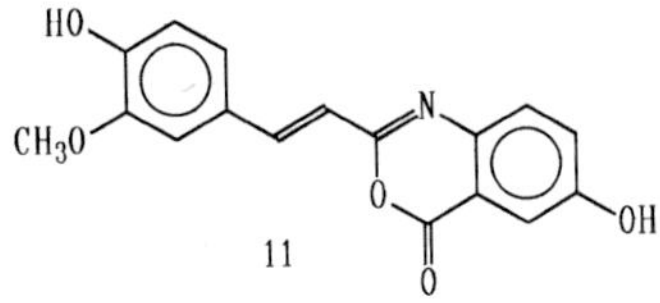

11

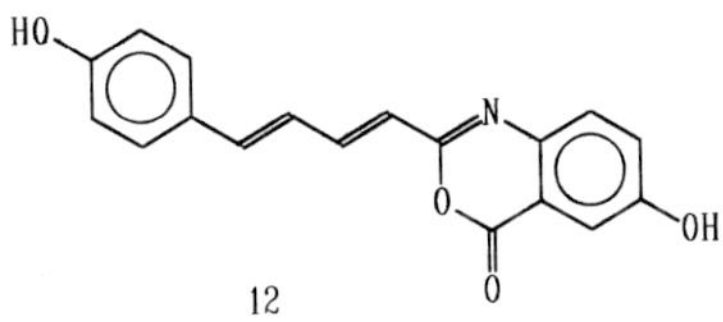

12

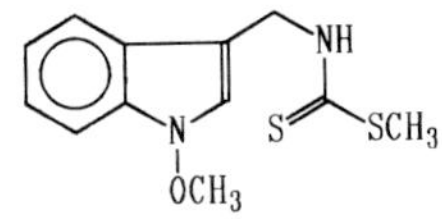

13

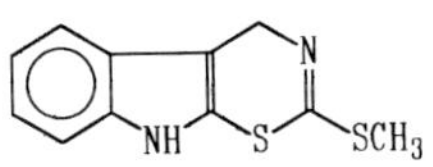

14

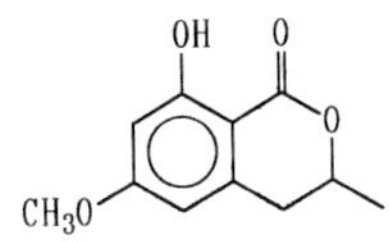

15

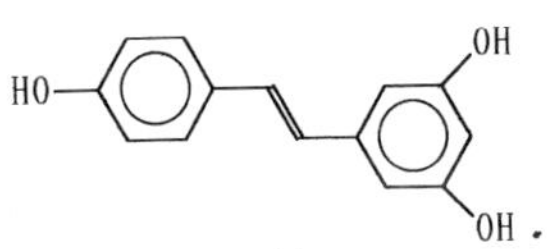

16

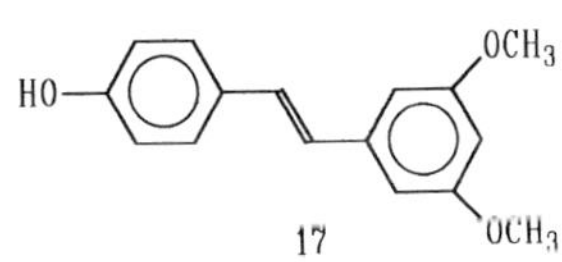

17

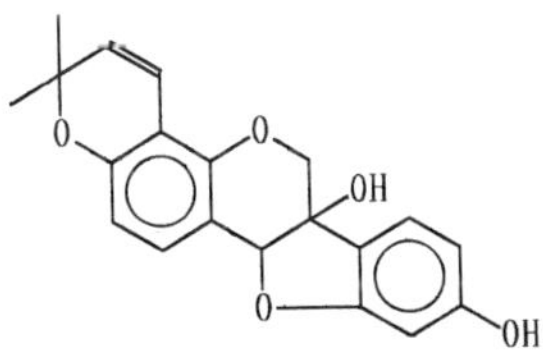

18

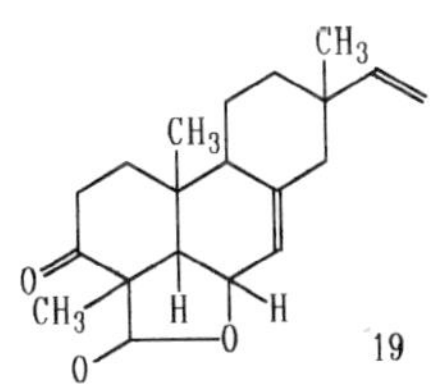

19

at 0.01–1.0 mM concentrations (Dron *et al.*, 1988; Wingate *et al.*, 1988) but so far the phenomenon has not been explained (Edwards *et al.*, 1991). Perhaps not surprisingly in view of the roles in signalling played by calcium and adenosine 3′,5′-cyclic monophosphate (cAMP) in a host of biological systems, they have also been invoked in the elicitation of the phytoalexin response (Kurosaki *et al.*, 1987; Zook *et al.*, 1987).

4.2.3 *Mechanism of expression of HR*

Early indications that a cell is about to die include membrane depolarization, electrolyte leakage and failure to plasmolyse normally in hypertonic solutions (reviewed by Slusarenko *et al.*, 1991). These symptoms all indicate membrane dysfunction. At present three main processes are reported to be instrumental in causing the membrane damage associated with the hypersensitive response:

(1) the production of active oxygen species such as singlet oxygen ($^1O_2$), the superoxide anion ($O_2^{\cdot -}$), the hydroxyl radical ($OH^{\cdot}$) and the perhydroxyl radical ($HO_2^{\cdot}$),
(2) the peroxidation of membrane lipids,
(3) the increased activity of the membranolytic enzymes, lipolytic acyl hydrolase and lipoxygenase.

Some of the more recent evidence for the operation of these processes follows.

(1) Anderson *et al.* (1991), using a galactoglucomannan elicitor from the α race of *Colletotrichum lindemuthianum* and suspension cultures of *Phaseolus vulgaris*, found that an immediate response of the cultured cells to the elicitor was the production of active oxygen species as determined by luminol-mediated chemiluminescence. The response peaked at 30 min.
(2) Croft *et al.* (1990) found that the evolution of ethane *in vivo*, an indicator of membrane peroxidation, occurred as a result of challenge of *P. vulgaris* with an incompatible but not a compatible race of *Ps. syringae* pv. *phaseolicola*.
(3) Croft *et al.* (1990) also found that lipolytic acyl hydrolase and lipoxygenase increased in leaves of *P. vulgaris* reacting hypersensitively to an incompatible race of *Ps. syringae* pv. *phaseolicola*. Moreover, when the induction of lipoxygenase was prevented by the protein synthesis inhibitor cycloheximide, hypersensitive necrosis to the incompatible race of bacterium was also prevented. In rather similar work with rice, Ohta *et al.* (1991) showed that lipoxygenase activity increased after infection with the rice blast fungus, *Magnaporthe grisea*. The increases were greater in plants inoculated with an incompatible race of the fungus compared with a compatible one.

Taking these results together, an early event in the hypersensitive response appears to be the production of active oxygen species. These would be expected to cause lipid peroxidation, resulting in membrane dysfunction (see Section 4.2.2). A second mechanism by which membranes may be attacked is through the operation of the enzymes lipolytic acyl hydrolase and lipoxygenase. Lipolytic acyl hydrolase releases fatty acids from membrane lipids and these are then subject to oxidation by lipoxygenase. Lipoxygenase itself is also reported to generate singlet oxygen (Siedow, 1991; Slusarenko *et al.*, 1991).

4.3 Phytoalexins

4.3.1 *Chemistry*

Phytoalexins belong to several classes of chemicals. Plants of the Leguminosae synthesize a large range of flavonoid compounds and, less commonly, stilbenes and polyacetylenes. Members of the Solanaceae, on the other hand, synthesize terpenoids. Nitrogen-containing compounds, the avenalumins, have been reported from oats and a sulphur-containing compound from Brassicas. Representative structures are given in Figure 4.2.

4.3.2 *Biochemical evidence for a role for phytoalexins in resistance*

Phytoalexins are biocides, affecting bacteria, fungi, nematodes, higher animals and plants (Smith and Banks, 1986). In addition, some flavonoid phytoalexins from legumes are oestrogenic. Despite the broad spectrum of activity of phytoalexins, organisms can vary widely in their sensitivity to them. For example, generally, Gram-positive bacteria are sensitive but Gram-negative bacteria less so. More extreme examples are found among strains of parasites which attack plants that accumulate phytoalexins, virulent strains generally tolerating higher concentrations than avirulent ones. Where this difference has been investigated it has usually been attributed to the ability of the virulent strains to degrade the phytoalexin.

Phytoalexins are non-specific, multi-site toxicants. When fungi are treated with phytoalexins their cytoplasm rapidly becomes granular and cell contents are lost, suggesting that membranes are one site of attack. For example, Amin *et al.* (1988) found that the carrot phytoalexin, 6-methoxymellein (Figure 4.2), interacted with the membranes of *Candida albicans* and caused leakage of cell contents. They also obtained evidence for the non-specificity of this phenomenon by making liposomes from phosphatidylcholine, dicetyl phosphate and cholesterol and monitoring permeability changes when they were treated with phytoalexin. Liposomes containing [^{14}C]-glucose, for example, released the labelled sugar following treatment with the phytoalexin.

Hoos and Blaich (1990) have investigated the effect of resveratrol (Figure 4.2), one of the phytoalexins synthesized by grape vines, on the germination of conidia and mycelial growth of *Botrytis cinerea*. They found that, when spores of *B. cinerea* were distributed sparsely over the surface of agar in which the phytoalexin had been incorporated, the germ tubes were malformed and ceased to grow. With higher densities, growth was apparently normal but large numbers of sclerotia, indicative of stress, were formed after 14 days. Pterostilbene is another stilbene phytoalexin synthesized by members of the Vitaceae (Figure 4.2). This compound caused the rapid destruction of ribosomes, endoplasmic reticulum and mitochondrial and nuclear membranes of *B. cinerea*, and within 3 h the plasma membrane was completely disorganized (Pezet and Pont, 1990).

Various criteria have been proposed to establish whether a phytoalexin can be considered to play a role in defence; the following are the most important.

(1) The compound must accumulate in response to infection.
(2) The compound must be inhibitory to the invading organism.
(3) The compound must accumulate to inhibitory concentrations in the vicinity of the parasite at the time the parasite ceases growing.
(4) Varying the rate of accumulation of the phytoalexin causes a corresponding variation in the resistance of the plant.
(5) Varying the sensitivity of the invading organism causes a corresponding variation in its virulence.

Of these criteria, the first three are mandatory and the remaining two provide corroborative evidence. Numerous studies have shown that phytoalexins accumulate

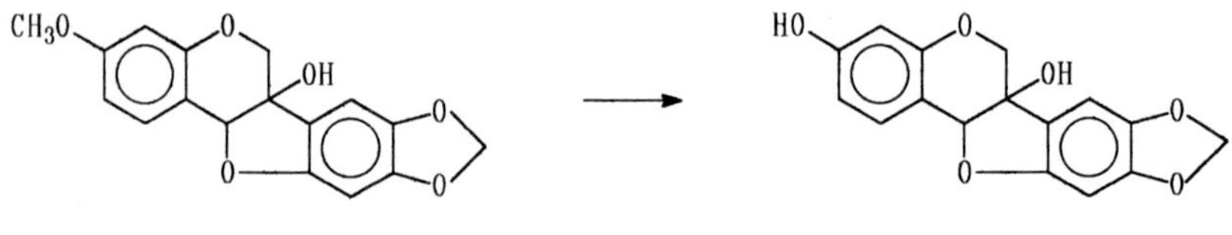

Figure 4.3. *Detoxification of the phytoalexin pisatin by demethylation.*

in response to infection and that the compounds are inhibitory to the invading organism. A more difficult task is to show that the phytoalexin reaches concentrations that are inhibitory to the parasite at the time it ceases to grow. This generally necessitates a sensitive means of phytoalexin detection within the challenged tissue. Using a radioimmunoassay, Hahn *et al.* (1985) were able to show that the concentration of the phytoalexin, glyceollin I (Figure 4.2), necessary to cause 90% inhibition of radial growth of the fungus *in vitro* was exceeded in resistant but not in susceptible soybean roots 8 h after inoculation.

Although many studies report the accumulation of phytoalexins in response to challenge of plants by fungi or bacteria, few have determined the response to nematodes. Recently, Huang and Barker (1991) examined the production of the soybean phytoalexin glyceollin I (Figure 4.2) when a resistant cultivar, Centennial, and a susceptible cultivar, Ransom, were inoculated by the soybean cyst nematode, *Heterodera glycines.* Using a radioimmunoassay, they found glyceollin I in the head region of the nematode 8 h after inoculation in the resistant cultivar and the concentration increased steadily to 0.3 $\mu mol\,cm^{-3}$ by 24 h. In contrast, none was found in the susceptible cultivar.

An early observation suggested that there was a correlation between phytoalexin tolerance and host range. Cruickshank (1962) found that of 50 fungal strains representing 45 species only five were tolerant of the pea phytoalexin, pisatin, and all were pathogens of pea. Of the remaining 45 strains, only one was a pea pathogen. Although subsequent work has shown that there are many exceptions, the concept that tolerance to a phytoalexin might be important for pathogenicity remains an important possibility (see Section 4.3.3).

4.3.3 *Genetic evidence for a role for phytoalexins in resistance*

Some of the most telling evidence for the role of phytoalexins in resistance comes from the laboratory of VanEtten (reviewed in VanEtten *et al.* 1989). Using pea and the fungus *Nectria haematococca,* they found that only those strains which were virulent to the plant were tolerant of its phytoalexin, pisatin, and could degrade it by demethylation (Figure 4.3). Moreover, when these were crossed with strains that were avirulent and sensitive to pisatin, the progeny generally fell into the two parental types, i.e. (i) those that were tolerant of pisatin, degraded it and were virulent to pea and (ii) those that were sensitive to pisatin, could not degrade it and were avirulent.

A more detailed analysis revealed that *N. haematococca* possesses multiple genes that confer characteristic rates of pisatin demethylation (*Pda* genes). So far, six genes have been identified but no field isolate has yet been found with more than two. The phenotypes conferred by these genes fall into three classes according to their ability to demethylate pisatin; those with a short lag in the production of pisatin demethylase and high activity of the enzyme (Pda^{SH}), those with a short lag and moderate activity (Pda^{SM}) and those with a long lag and low activity (Pda^{LL}). Genetic analysis showed that only Pda^{SH} and Pda^{SM} were virulent while Pda^{LL} phenotypes were no more virulent than Pda^{-}. Moreover, in a cross between Pda^{SH} and Pda^{SM} the relative virulence of the progeny was strongly correlated with their level of inducible pisatin demethylase activity.

The importance of pisatin demethylase has been corroborated by experiments in which a Pda^{-} isolate was transformed with a fragment of DNA containing a gene which conferred the ability to demethylate pisatin (Ciufetti *et al.*, 1988). Three of the transformants were more tolerant of pisatin and two of these were significantly more pathogenic to pea.

Recently, VanEtten and co-workers (Miao *et al.*, 1991) have mapped one *Pda* gene (*Pda6*) to a small meiotically unstable chromosome which was dispensable for normal growth. They point out that these properties are characteristic of B chromosomes and suggest that such genetic elements could be an important way by which variation could be generated in pathogenic fungi. This would be of particular significance if such genetic elements were laterally transmissible and could confer phytoalexin tolerance to other fungi in a manner analogous to R plasmids in bacteria (Miao *et al.*, 1991).

4.4 Conclusions

The occurrence of cell death in response to challenge by obligate parasites provides a logical reason for the role of the HR in resistance. Since the host cells are killed and the parasite, as an obligate, needs live cells in order to establish a parasitic relationship, it too must die.

However, the HR is also associated with resistance to facultative parasites where host cell death *per se* is not a sufficient reason for resistance. In these, one question that arises is whether the HR is a prerequisite for resistance or unconnected with it. Certainly defence responses that are not mediated through the HR are known. For example, Voisey and Slusarenko (1989) found that heat- or UV-killed cells of both avirulent and virulent races of *Ps. syringae* pv. *phaseolicola* induced chitinase activity in bean leaves but not the HR or phytoalexin synthesis. On the other hand, active oxygen species are associated with the HR and they are non-specific toxins. Thus, one way of regarding the death of host cells in the HR is as the inevitable consequence of the method used to kill pathogen cells, i.e. the production of active oxygen species.

Recently, Li *et al.* (1991) have suggested a mechanism by which the hypersensitive response and phytoalexin production may be linked. They found that the 13-hydroperoxides and 13-hydroxides of linoleic and linolenic acids increased rapidly in rice leaves inoculated with the blast pathogen. These compounds were not only anti-fungal themselves but also induced the production of the rice phytoalexin, momilactone A (Figure 4.2).

Perhaps the most exciting prospect of all arises from the work of Kearney and Staskawicz (1990). They have shown that the *avrBs2* gene of the pepper pathogen, *Xanthomonas campestris* pv. *vesicatoria,* is also present in several other pathovars which infect different hosts. As stated previously (Section 4.2.1), this gene is a necessary fitness component of the bacterium and as a result the corresponding gene in the plant which confers the HR, *Bs2,* has given resistance that has proved durable. Thus, once the *Bs2* gene is isolated and cloned it could be introduced by genetic engineering into the hosts of other *X. campestris* pathovars where it would similarly be expected to confer durable resistance.

References

Amin, M., Kurosaki, F. and Nishi, A. (1986) Extracellular pectinolytic enzymes of fungi elicit phytoalexin accumulation in carrot suspension culture. *J. Gen. Microbiol.* **132**, 771–777.

Amin, M., Kurosaki, F. and Nishi, A. (1988) Carrot phytoalexin alters the membrane permeability of *Candida albicans* and multilamellar liposomes. *J. Gen. Microbiol.* **134**, 241–246.

Anderson, A.J., Rogers, K., Tepper, C.S., Blee, K. and Cardon, J. (1991) Timing of molecular events following elicitor treatment of plant cells. *Physiol. Mol. Plant Pathol.* **38**, 1–13.

Arlat, M., Gough, C.L., Barber, C.E., Boucher, C. and Daniels, M.J. (1991) *Xanthomonas campestris* contains a cluster of HRP genes related to the larger HRP cluster of *Pseudomonas solanacearum. Mol. Plant Microb. Interact.* **4**, 593–601.

Bailey, J.A. (1982) Mechanism of phytoalexin accumulation. In: *Phytoalexins* (ed. J.A. Bailey and J.W. Mansfield). Blackie, Glasgow, pp. 289–318.

Barber, M.S., Bertram, R.E. and Ride, J.P. (1989) Chitin oligosaccharides elicit lignification in wounded wheat leaves. *Physiol. Plant Pathol.* **34**, 3–12.

Bauer, D.W. and Beer, S.V. (1991) Further characterization of an *hrp* gene cluster of *Erwinia amylovora. Mol. Plant Microb. Interact.* **4**, 493–499.

Bonas, U., Stall, R.E. and Staskawicz, B. (1989) Genetic and structural characterization of the avirulence gene *avrB3* from *Xanthomonas campestris* pv. *vesicatoria. Mol. Gen. Genet.* **218**, 127–136.

Bostock, R.M., Kuć, J. and Laine, R.A. (1981) Eicosapentaenoic and arachidonic acids from *Phytophthora infestans* elicit fungitoxic sesquiterpenes in potato. *Science* **212**, 67–69.

Canteros, B., Minsavage, G., Bonas, U., Pring, D. and Stall, R. (1991) A gene from *Xanthomonas campestris* pv. *vesicatoria* that determines avirulence in tomato is related to *avrBs3. Mol. Plant Microb. Interact.* **4**, 628–632.

Cheong, J.J., Birberg, W., Fugedi, P., Pilotti, A., Garegg, P.J., Hong, N., Ogawa, T. and Hahn, M.G. (1991) Structure-activity relationships of oligo-beta-glucoside elicitors of phytoalexin accumulation in soybean. *Plant Cell* **3**, 127–136.

Cheong, J.J. and Hahn, M.G. (1991) A specific, high-affinity binding site for the hepta-beta-glucoside elicitor exists in soybean membranes. *Plant Cell* **3**, 137–147.

Ciufetti, L.M., Welltring, K.-M., Turgeon, B.G., Yoder, O.C. and VanEtten, H.D. (1988) Transformation of *Nectria haematococca* with a gene for pisatin demethylating activity, and the role of pisatin detoxification in virulence. *J. Cell Biochem.* **12C**, 278 (Abstr.).

Cooksey, C.J., Garratt, P.J., Dahiya, J.S. and Strange, R.N. (1983) Sucrose: a constitutive elicitor of phytoalexin synthesis. *Science* **220**, 1398–1400.

Croft, K.P.C., Voisey, C.R. and Slusarenko, A.J. (1990) Mechanism of hypersensitive cell collapse – correlation of increased lipoxygenase activity with membrane damage in leaves of *Phaseolus vulgaris* (L) inoculated with an avirulent race of *Pseudomonas syringae* pv. *phaseolicola. Physiol. Mol. Plant Pathol.* **36**, 49–62.

Cruickshank, I.A.M. (1962) Studies on phytoalexins. IV. The antimicrobial spectrum of pisatin. *Aust. J. Biol. Sci.* **15**, 147–159.

Cruickshank, I.A.M. and Perrin, D.R. (1968) The isolation and partial characterization of monilicolin A, a polypeptide with phaseollin-inducing activity from *Monilinia fructicola. Life Sci.* **7**, 449–458.

Culver, J.N. and Dawson, W.O. (1991) Tobacco mosaic virus elicitor coat protein genes produce a hypersensitive phenotype in transgenic *Nicotiana sylvestris* plants. *Mol. Plant Microb. Interact.* **4**, 458–463.

Davis, K.R., Darvill, A.G., Albersheim, P. and Dell, A. (1986) Host-pathogen interactions. XXIX. Oligogalacturonides released from sodium polypectate by endopolygalacturonic acid lyase are elicitors of phytoalexins in soybean. *Plant Physiol.* **80**, 568–577.

De Feyter, R. and Gabriel, D.W. (1991) At least 6 avirulence genes are clustered on a 90-kilobase plasmid in *Xanthomonas campestris* pv. *malvacearum. Mol. Plant Microb. Interact.* **4**, 423–432.

Dron, M., Clouse, S.D., Dixon, R.A., Lawton, M.A. and Lamb, C.J. (1988) Glutathione and fungal elicitor regulation of a plant defense gene promoter in electroporated protoplasts. *Proc. Natl. Acad. Sci. USA* **85**, 6738–6742.

Edwards, R., Blount, J.W. and Dixon, R.A. (1991) Glutathione and elicitation of the phytoalexin response in legume cell cultures. *Planta* **184**, 403–409.

Epperlein, M.M., Noronha-Dutra, A.A. and Strange, R.N. (1986) Involvement of the hydroxyl radical in the abiotic elicitation of phytoalexins in legumes. *Physiol. Mol. Plant Pathol.* **28**, 67–77.

Flor, H.H. (1971) Current status of the gene-for-gene concept. *Ann. Rev. Phytopathol.* **9**, 275–296.

Hahn, M.G., Bonhoff, A. and Grisebach, H. (1985) Quantitative localization of the phytoalexin glyceollin I in relation to fungal hyphae in soybean roots infected with *Phytophthora megasperma* f. sp. *glycinea. Plant Physiol.* **77**, 591–601.

Hoos, G. and Blaich, R. (1990) Influence of resveratrol on germination of conidia and mycelial growth of *Botrytis cinerea* and *Phomopsis viticola. J. Phytopathol.* **129**, 102–110.

Huang, H.C., Hutcheson, S.W. and Collmer, A. (1991) Characterization of the *hrp* cluster from *Pseudomonas syringae* pv. *syringae*-61 and TnphoA tagging of genes encoding exported or membrane-spanning hrp proteins. *Mol. Plant Microb. Interact.* **4**, 469–476.

Huang, H.C., Schuurink, R., Denny, T.P., Atkinson, M.M., Baker, C.J., Yucel, I., Hutcheson, S.W. and Collmer, A. (1988) Molecular cloning of a *Pseudomonas syringae* pv. *syringae* gene cluster that enables *Pseudomonas fluorescens* to elicit the hypersensitive response in tobacco. *J. Bacteriol.* **170**, 4748–4756.

Huang, J.S. and Barker, K.R. (1991) Glyceollin-I in soybean-cyst nematode interactions – spatial and temporal distribution in roots of resistant and susceptible soybeans. *Plant Physiol.* **96**, 1302–1307.

Jenner, C., Hitchin, E., Mansfield, J., Walters, K., Betteridge, P., Teverson, D. and Taylor, J. (1991) Gene-for-gene interactions between *Pseudomonas syringae* pv. *phaseolicola* and *Phaseolus. Mol. Plant Microb. Interact.* **4**, 553–562.

Kearney, B., Ronald, P.C., Dahlbeck, D. and Staskawicz, B.J. (1988) Molecular basis for evasion of plant host defence in bacterial spot disease of pepper. *Nature* **332**, 541–543.

Kearney, B. and Staskawicz, B.J. (1990) Widespread distribution and fitness contribution of *Xanthomonas campestris* avirulence gene *avrbs2. Nature* **346**, 385–386.

Keen, N.T. (1990) Gene-for-gene complementarity in plant pathogen interactions. *Ann. Rev. Genet.* **24**, 447–463.

Keen, N.T., Tamaki, S., Kobayashi, D., Gerhold, D., Stayton, M., Shen, H., Gold, S., Lorang, J., Thordal-Christensen, H., Dahlbeck, D. and Staskawicz, B.J. (1990) Bacteria expressing avirulence gene D produce a specific elicitor of the soybean hypersensitive reaction. *Mol. Plant Microb. Interact.* **3**, 112–121.

Kobayashi, D.Y., Tamaki, S.J. and Keen, N.T. (1989) Cloned avirulence genes from the tomato pathogen *Pseudomonas syringae* pv. *tomato* confer cultivar specificity on soybean. *Proc. Natl. Acad. Sci. USA* **86**, 157–161.

Kogel, G., Beissmann, B., Reisener, H.J. and Kogel, K.H. (1988) A single glycoprotein from *Puccinia graminis* f.sp. *tritici* cell walls elicits the hypersensitive lignification response in wheat. *Physiol. Mol. Plant Pathol.* **33**, 173–185.

Kogel, G., Beissmann, B., Reisener, H.J. and Kogel, K. (1991) Specific binding of a hypersensitive lignification elicitor from *Puccinia graminis* f. sp. *tritici* to the plasma membrane from wheat (*Triticum aestivum* L). *Planta*, **183**, 164–169.

Kurosaki, F., Tsurusawa, Y. and Nishi, A. (1987) The elicitation of phytoalexins by Ca^{2+} and cyclic AMP in carrot cells. *Phytochem.* **26**, 1919–1923.

Lee, S.C. and West, C.A. (1981a) Polygalacturonase from *Rhizopus stolonifer*, an elicitor of casbene synthetase activity in castor bean (*Ricinus communis* L.) seedlings. *Plant Physiol.* **67**, 633–639.

Lee, S.C. and West, C.A. (1981b) Properties of *Rhizopus stolonifer* polygalacturonase, an elicitor of casbene synthetase activity in castor bean (*Ricinus communis* L.) seedlings. *Plant Physiol.* **67**, 640–645.

Li, W.X., Kodama, O. and Akatsuka, T. (1991) Role of oxygenated fatty acids in rice phytoalexin production. *Agric. Biol. Chem.* **55**, 1041–1047.

Lindgren, P.B., Panopoulos, N.J., Staskawicz, B.J. and Dahlbeck, D. (1988) Genes required for pathogenicity and hypersensitivity are conserved and interchangeable among pathovars of *Pseudomonas syringae*. *Mol. Gen. Genet.* **211**, 499–506.

Lindgren, P.B., Peet, R. and Panopoulos, N.J. (1986) Gene cluster of *Pseudomonas syringae* pv. "*phaseolicola*" controls pathogenicity on bean plants and hypersensitivity on non-host plants. *J. Bacteriol.* **168**, 512–522.

Malamy, J., Carr, J.P., Klessig, D.F. and Raskin, I. (1990) Salicylic acid – a likely endogenous signal in the resistance response of tobacco to viral infection. *Science* **250**, 1002–1004.

Maniara, G., Laine, R. and Kuć, J. (1984) Oligosaccharides from *Phytophthora infestans* enhance the elicitation of sesquiterpenoid stress metabolites by arachidonic acid in potato. *Physiol. Plant Pathol.* **24**, 177–186.

Métraux, J.P., Signer, H., Ryals, J., Ward, E., Wyssbenz, M., Gaudin, J., Raschdorf, K., Schmid, E., Blum, W. and Inverardi, B. (1990) Increase in salicylic acid at the onset of systemic acquired resistance in cucumber. *Science* **250**, 1004–1006.

Miao, V.P., Covert, S.F. and VanEtten, H.D. (1991) A fungal gene for antibiotic resistance on a dispensable (B) chromosome. *Science* **254**, 1773–1776.

Napoli, C. and Staskawicz, B. (1987) Molecular characterization and nucleic acid sequence of an avirulence gene from race 6 of *Pseudomonas syringae* pv. *glycinea*. *J. Bacteriol.* **169**, 572–578.

Ohta, H., Shida, K., Peng, Y.L., Furusawa, I., Shishiyama, J., Aibara, S. and Morita, Y.H. (1991) A lipoxygenase pathway is activated in rice after infection with the rice blast fungus *Magnaporthe grisea*. *Plant Physiol.* **97**, 94–98.

Paxton, J. (1981) Phytoalexins – a working redefinition. *Phytopathologische Z.* **101**, 106–109.

Pezet, R. and Pont, V. (1990) Ultrastructural observations of pterostilbene fungitoxicity in dormant conidia of *Botrytis cinerea* Pers. *J. Phytopathol.* **129**, 19–30.

Rahme, L.G., Mindrinos, M.N. and Panapoulos, N.J. (1991) Genetic and transcriptional organization of the *hrp* cluster of *Pseudomonas syringae* pv. *phaseolicola*. *J. Bacteriol.* **173**, 575–586.

Rasmussen, J.B., Hammerschmidt, R. and Zook, M.N. (1991) Systemic induction of salicylic acid accumulation in cucumber after inoculation with *Pseudomonas syringae* pv. *syringae*. *Plant Physiol.* **97**, 1342–1347.

Ronald, P.C. and Staskawicz, B.J. (1988) The avirulence gene *avrBs1* from *Xanthomonas campestris* pv. *vesicatoria* encodes a 50-kD protein. *Mol. Plant Microb. Interact.* **1**, 191–198.

Schottens-Toma, I.M.J. and DeWit, P.J.G.M. (1988) Purification and primary structure of a necrosis-inducing peptide from the apoplastic fluids of tomato infected with *Cladosporium fulvum* (Syn. *Fulvia fulva*). *Physiol. Plant Pathol.* **33**, 59–67.

Siedow, J.N. (1991) Plant lipoxygenase – structure and function. *Ann. Rev. Plant Physiol. Plant Mol. Biol.* **42**, 145–188.

Slusarenko, A.J., Croft, K.P. and Voisey, C.R. (1991) Biochemical and molecular events in the hypersensitive response of bean to *Pseudomonas phaseolicola* pv. *phaseolicola*. In: *Biochemistry and Molecular Biology of Plant-Pathogen Interactions* (ed. C.J. Smith). Clarendon Press, Oxford, pp. 126–143.

Smith, D.A. and Banks, S.W. (1986) Biosynthesis, elicitation and biological activity of isoflavonoid phytoalexins. *Phytochem.* **25**, 979–995.

Staskawicz, B., Dahlbeck, D. and Keen, N.T. (1984) Cloned avirulence gene of *Pseudomonas syringae* pv. *glycinea* determines race-specific incompatibility on *Glycine max*. *Proc. Natl. Acad. Sci. USA*, **81**, 6024–6028.

Tamaki, S., Dahlbeck, D., Staskawicz, B.J. and Keen, N.T. (1988) Characterization and expression of two avirulence genes cloned from *Pseudomonas syringae* pv. *glycinea*. *J. Bacteriol.* **170**, 4846–4854.

Tsukaya, H., Ohshima, T., Naito, S., Chino, M. and Komeda, Y. (1991) Sugar-dependent expression of the *CHS-A* gene for chalcone synthase from *Petunia* in transgenic *Arabidopsis*. *Plant Physiol.* **97**, 1414–1421.

VanEtten, H.D., Matthews, D.E. and Matthews, P.S. (1989) Phytoalexin detoxification: importance for pathogenicity and practical implications. *Ann. Rev. Phytopathol.* **27**, 143–164.

Van Kan, J.A.L., van den Ackerveken, G.F.J.M. and de Wit, P.J.G.M. (1991) Cloning and characterization of cDNA of avirulence gene *avr9* of the fungal pathogen *Cladosporium fulvum*, causal agent of tomato leaf mold. *Mol. Plant Microb. Interact.* **4**, 52–59.

Voisey, C.R. and Slusarenko, A.J. (1989) Chitinase mRNA and enzyme activity in *Phaseolus vulgaris* (L.) increase more rapidly in response to avirulent than to virulent cells of *Pseudomonas syringae* pv. *phaseolicola*. *Physiol. Mol. Plant Pathol.* **35**, 403–412.

Ward, E.R., Uknes, S.J., Williams, S.C., Dincher, S.S., Wiederhold, D.L., Alexander, D.C., Ahlgoy, P., Métraux, J.P. and Ryals, J.A. (1991) Coordinate gene activity in response to agents that induce systemic acquired resistance. *Plant Cell* **3**, 1085–1094.

Willis, D.K., Rich, J.J. and Hrabak, E.M. (1991) *hrp* genes of phytopathogenic bacteria. *Mol. Plant Microb. Interact.* **4**, 132–138.

Wingate, V.P.M., Lawton, M.A. and Lamb, C.J. (1988) Glutathione causes a massive and selective induction of plant defense genes. *Plant Physiol.* **87**, 206–210.

Zook, M.N., Rush, J.S. and Kuć, J.A. (1987) A role for Ca^{2+} in the elicitation of rishitin and lubimin accumulation in potato tissue. *Plant Physiol.* **84**, 520–525.

5

Altered patterns of protein synthesis: their significance in defence and growth

M.T. Esquerré-Tugayé, J. Fournier and M. Rickauer

5.1 Introduction

It has long been known that protein synthesis in plants is profoundly altered upon infection by pathogens, and in response to various stresses. Changes in patterns of proteins can best be visualized by two-dimensional gel electrophoresis of the products obtained from *in vitro* translation of mRNA (Figure 5.1), when both positive and negative changes can be observed (Roby, 1987). Very often it is difficult to assign functions to most of these changes in proteins, particularly when they affect discrete spots on the gels. In addition, patterns obtained from infected tissues are complicated because they include both host and pathogen proteins.

A proper appraisal of plant responses requires the use of model systems, i.e. systems in which pathogens are replaced by their elicitors, or interactions whose specificity allows protein changes to be clearly assigned to the host plant. Differential screening of cDNA libraries, comparing induced and non-induced material, may thus lead to the recovery of cDNA clones corresponding to mRNAs specific to a particular situation. Such an approach can eventually result in the identification of gene products and their functions. In practice, we have taken a reverse, complementary approach. On the basis of our present knowledge of plant–micro-organism interactions, proteins have been identified whose functions are likely to be involved in the cascade leading from recognition to host defence. This, in turn, has enabled identification of the corresponding genes, and gene expression studies. The role of proteins in early events related to signalling phenomena, active defence and certain critical metabolic alterations following injury are considered here.

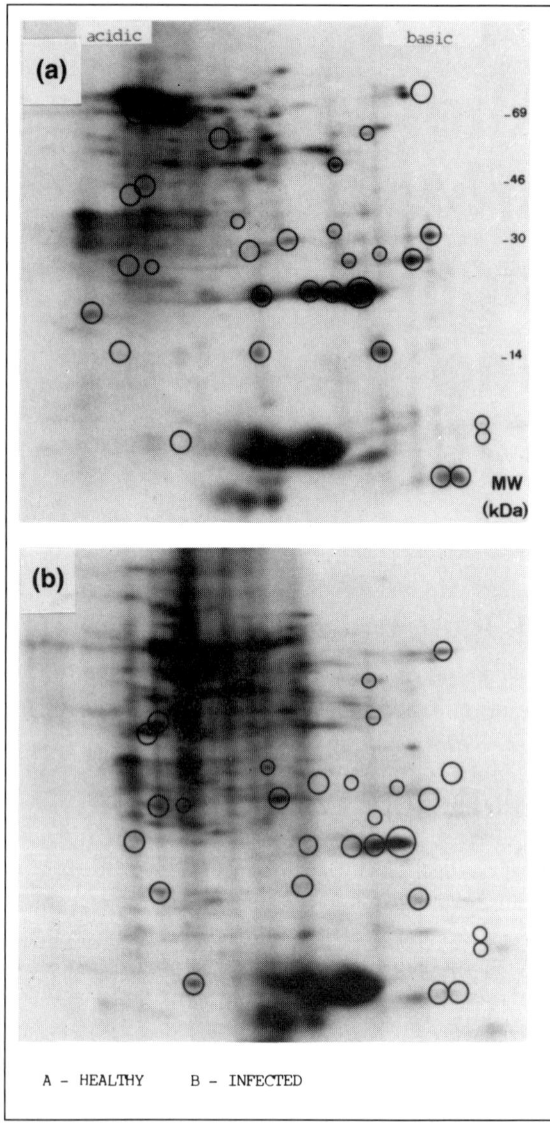

Figure 5.1. *Two-dimensional gel electrophoresis of the peptides recovered after* in vitro *translation of poly A+RNA isolated from* Colletotrichum lagenarium *infected melon seedlings (b) at the time of symptom appearance, and their healthy controls (a).*

5.2 Early events and associated proteins

Cell surfaces are of prime importance in the outcome of plant–pathogen interactions (Heath, 1986). The hypothesis that early events take place at the cell wall–plasmalemma interface received full support when it was demonstrated that membrane depolarization occurs in host plants upon infection or elicitor treatment (Katou *et al.*, 1982; Pélissier *et al.*, 1986). This effect was recorded in the very first seconds following

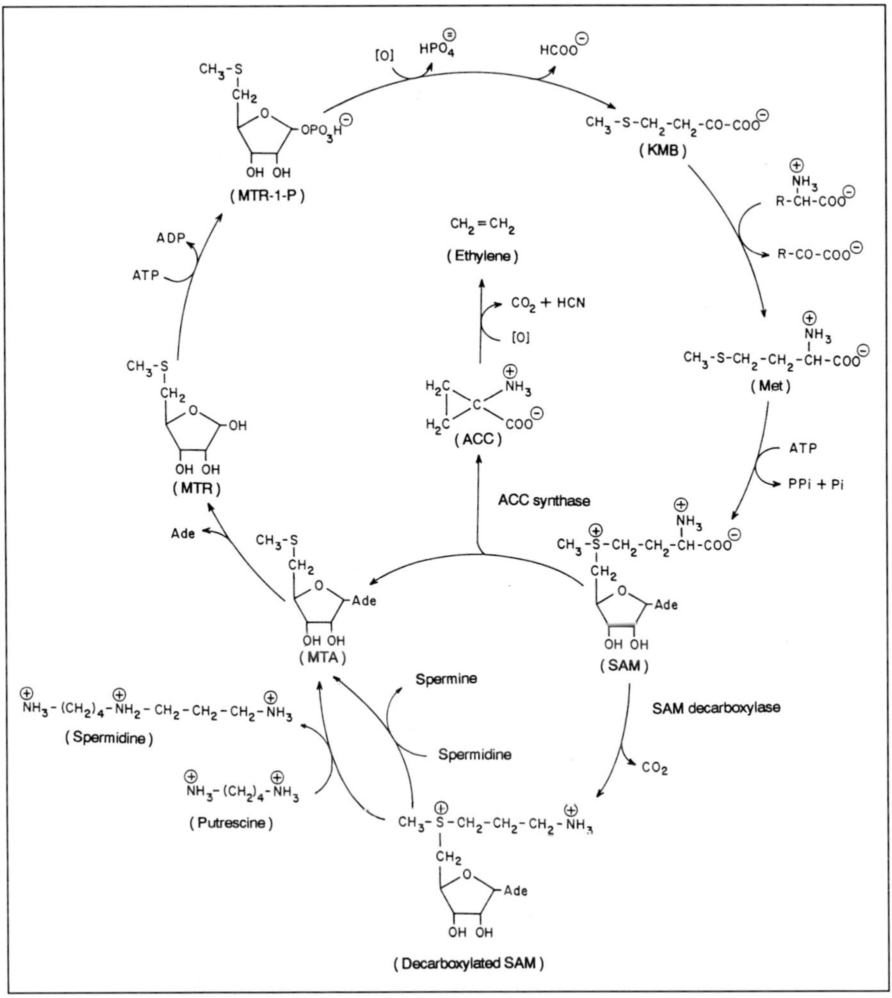

Figure 5.2. *Methionine cycle in relation to polyamine and ethylene biosynthesis. Abbreviations: ACC, 1-aminocyclopropane-1-carboxylic acid; Ade, adenine; KMB, 2-keto-4-methylthiobutyric acid; Met, methionine; MTA, 5′-methylthioadenosine; MTR, 5-methylthioribose; MTR-1-P, 5-methylthioribose 1-phosphate; SAM, S-adenosylmethionine. (Modified from Miyazaki and Yang, 1987).*

addition of the elicitor of *Phytophthora megasperma* f.sp. *glycinea* to cotyledonary cells of soybean (Mayer and Ziegler, 1988), which is the most rapid response registered so far. This observation suggests that, in addition to being the first step of a putative transduction cascade, changes in membrane potential will result in alterations in a suite of membrane-related functions. Ethylene biosynthesis and phospholipid breakdown might represent two of them. The methionine pathway of ethylene biosynthesis (Figure 5.2) is highly and transiently stimulated in infected plants and in response to elicitors (Toppan and Esquerré-Tugayé, 1984; Rickauer *et al.*, 1990). The burst of ethylene production results from a previous increase of ACC-synthase activity and of ACC (1-aminocyclopropane-1-carboxylic acid) the direct precursor of the hormone (Rickauer *et al.*, 1990). It has been shown in parsley and in tobacco cells that transcription is not required for the triggering of ACC-synthase (Chappell *et al.*, 1984; Rickauer *et al.*, 1990). Whether ethylene is subsequently involved in the numerous protein changes recorded after infection is still an open question. *In vitro* translation of mRNA isolated from ethylene-treated tissues shows several modifications in the pattern of proteins, but the effect of ethylene synthesized *in vivo* might be quite different. Doubts about the importance of ethylene are strengthened by two observations. Firstly, activity of S-adenosyl methionine (SAM)-synthase, an enzyme involved in polyamine and ethylene biosynthesis, is also enhanced in response to fungal elicitors (unpublished data). Polyamines are well known for their ability to regulate gene expression. Secondly, inhibition of ethylene biosynthesis by aminoethoxyvinylglycine (AVG) does not necessarily abolish other protein changes. This is notably the case for lipoxygenase.

Lipoxygenase activity is highly and transiently increased in the incompatible interaction between tobacco and *Phytophthora parasitica nicotianae*, and in a more delayed manner during a compatible interaction. A similar phenomenon has been reported in *Magnaporthe grisea*-infected rice seedlings (Ohta *et al.*, 1991). In tobacco cells, lipoxygenase is induced within the first 3 h following the addition of elicitors prepared from *Phytophthora* and shows maximum activity at about 24 h (Rickauer *et al.*, 1990). This enzyme catalyses the oxidation of polyunsaturated fatty acids, e.g. linolenic acid, by molecular oxygen, giving rise to the formation of hydroperoxides which can be further converted either chemically or enzymically into a series of compounds. One of them, jasmonic acid, has hormone-like activities. It was recently found that it elicits proteinase inhibitor gene expression in tomato and tobacco (Farmer and Ryan, 1990; Rickauer *et al.*, 1992), and that it induces numerous protein changes. It is striking to note that both ethylene biosynthesis and the lipoxygenase pathway are usually involved in senescence processes. As such, they might account for the localized cell death which typifies the hypersensitive response of resistant plants (see Chapter 4, this volume), and for the extended necrosis which occurs in susceptible plants. The development of a defence response is coincident with these processes.

5.3 Responses related to active defence

Besides enzymes involved in the biosynthesis of phytoalexins (see Chapter 4) and lignin, several classes of proteins participate in the defence of plants against pathogens (Collinge and Slusarenko, 1987). They can be divided into cell wall glycoproteins, hydrolases, and inhibitors of microbial enzymes. In only a very few systems have the kinetics of their induction, their localization and mode of action been thoroughly studied. The available data indicate that they share several features (Table 5.1), namely induction at the right time, right location, in sufficient amounts, and in response to elicitors. In addition, they are induced in plants regardless of the infective microbe, and in response to developmental and environmental factors. Changes in the three main classes of defence proteins are considered.

Table 5.1. *Common features of defence proteins*

1.	Early induction during race-cultivar specific resistance
2.	Induction by elicitors and possibly related signals (ethylene, jasmonic acid)
3.	Induction at sites of resistance (papillae, wall appositions...) during incompatible interactions
4.	Induction in infected plants regardless of the infective microbe
5.	Antagonistic effects against microbes due to growth inhibition and hydrolysis
6.	Induction under developmental and environmental conditions (hormones, light, temperature, water stress, anoxia, flowering, subculturing)

5.3.1 *Hydroxyproline-rich glycoproteins (HRGPs)*

Large increases in HRGP (extensin) were first demonstrated in fungal-infected melon seedlings (Esquerré-Tugayé, 1973; Esquerré-Tugayé and Mazau, 1974). It has been reported that intense deposition of HRGP is associated with the defence of plants against pathogens (Esquerré-Tugayé *et al.*, 1979; Mazau and Esquerré-Tugayé, 1986). In bean and melon infected by *Colletotrichum*, and in response to its elicitors, several transcripts encoding HRGPs (Showalter *et al.*, 1985; Rumeau *et al.*, 1987) and several products (Mazau *et al.*, 1988) have been identified. Starting from the low levels already present in the cell walls of healthy plants, amounts increase earlier in resistant than in susceptible cultivars, and in an organ-specific manner. Antibodies raised against the most abundant cell wall HRGP in melon allowed the detection of HRGP accumulation at sites of resistance to fungal and bacterial pathogens, and to viruses, notably in papillae, wall appositions and intercellular spaces (Benhamou *et al.*, 1990; O'Connell *et al.*, 1990; Benhamou *et al.*, 1991a). Increases in cell wall HRGPs have been found in all infected plants studied so far, with the exception of monocots (grasses), whether the plant is infected by a fungus, bacterium or virus. Recently, we demonstrated that HRGPs also accumulate at the interface between host plants and symbiotic micro-organisms (*Rhizobium* and certain endomycorrhizas) (Benhamou *et al.*, 1991b; Bonfante-Fasolo *et al.*, 1991). Although HRGPs are chemically better characterized than any other defence protein, their precise role is not totally understood. Due to their structure and polycationic nature, it is assumed that they strengthen the cell wall and agglutinate negatively charged micro-organisms.

HRGP genes are differentially transcribed under the control of developmental and environmental factors such as ethylene, wounding and flowering (Corbin *et al.*, 1987; Rumeau *et al.*, 1988; Neale *et al.*, 1990). Other cell wall glycoproteins, such as glycine-rich proteins, thionins, and proline-rich glycoproteins accumulate in response to external stimuli (Ye *et al.*, 1991), thus indicating the importance of cell surface protein changes in the adaptation of plants to their environment.

5.3.2 *β-1,3-glucanases and chitinases*

In contrast to the highly insoluble HRGPs, β-1,3-glucanases and chitinases are readily soluble and easy to identify in electrophoretic gels along with other pathogenesis-related proteins of unknown functions (Kauffmann *et al.*, 1987). They comprise several endo- and exo-enzymes, some of them being constitutive while others are synthesized *de novo*, and they accumulate either in extracellular spaces or within the vacuole. In the race-cultivar specific interaction between *Phaseolus vulgaris* and *Colletotrichum lindemuthianum*, they increase earlier in resistant than in susceptible cultivars, both at the gene and at the protein levels (Daugrois *et al.*, 1990). A detailed study of β-1,3-glucanase in this system indicated that, of the five enzymes endowed with this activity, two acidic endoforms (mol.wt 36.5 and 36 kDa) were highly induced (Daugrois *et al.*, 1992), and one was very active against *Colletotrichum* cell walls. As in other systems, the two acidic enzymes accumulated in the intercellular spaces, mainly in leaves.

Besides infection, it has been reported that elicitors, ethylene and flowering induce β-1,3-glucanases and chitinases (Broglie *et al.*, 1986; Roby and Esquerré-Tugayé, 1986; Neale *et al.*, 1990). Melon seedlings expressing a high level of chitinase activity in response to elicitor treatment have relatively high resistance to *Colletotrichum lagenarium* (Roby *et al.,* 1988b). Despite extended studies of these proteins at the molecular level, their enzymic properties (e.g. K_m, or V_{max}) are still unknown. The significance of their induction during viral infection is not understood, since viruses do not represent potential substrates.

5.3.3 *Peptide inhibitors of microbial enzymes*

Pectinases and proteases are important pathogenic products which can be counteracted by peptide inhibitors present in the cell wall or in the vacuole of the host. An endopolygalacturonase inhibitor (PGIP) has been purified from *Phaseolus vulgaris* (Lafitte *et al.*, 1984). It is constitutively present in the cell walls, and in higher amounts in cultivars resistant to *Colletotrichum* than in those susceptible. Its high specific affinity allowed endoPG (endopolygalacturonase) to be localized in infected tissues when gold-labelled PGIP was applied (Benhamou *et al.,* 1991c). It is assumed that PGIP plays an important role in signalling regulation, through the lowering of endoPG activity. This enzyme releases pectic fragments whose activity as elicitors of defence responses depends on their length (Cervone *et al.,* 1989; see also p. 36). By preventing extended degradation of pectic polysaccharides, PGIP activity would favour the appearance of active fragments.

Proteinase inhibitors (PIs) are low molecular weight compounds that have been found to accumulate in the vacuole of plants upon wounding (Ryan, 1990). Some proteinase inhibitors are present in seeds and storage organs, and their synthesis is under the control of abscisic acid (Pena-Cortès *et al.*, 1989). As in the case of PGIP, very few studies of proteinase inhibitors have been made on infected plants. However, in the melon–*Colletotrichum lagenarium* system, they are highly induced and exert activity against the proteases of the pathogen (Roby *et al.*, 1987). Recently, a study of *Phytophthora*-infected tobacco seedlings showed that amounts are higher in an incompatible interaction than in a compatible interaction. Interestingly, they were also induced by jasmonic acid, one of the by-products of lipoxygenase action (Rickauer *et al.*, 1992). In tomato and potato, PIs are encoded by a small multigene family (Graham *et al.*, 1986; Pena-Cortès *et al.*, 1989). Transgenic tobacco plants constitutively expressing a high level of PI have enhanced resistance to pest attack (Johnson *et al.*, 1989).

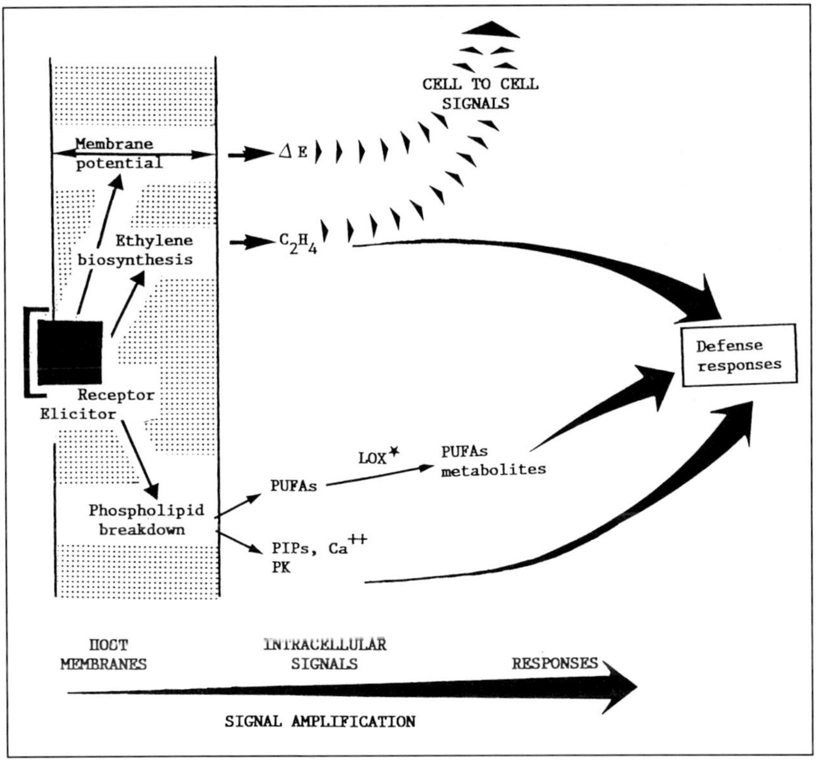

Figure 5.3. *Signal transduction in response to elicitors : an hypothesis. Abbreviations:* ΔE = *variations in membrane potential; PUFAs, polyunsaturated fatty acids; PIPs, phosphoinositides; PK, protein kinases; LOX, lipoxygenase.*

5.4 Metabolic alterations

One of the best identified metabolic alterations following infection consists of a dramatic decrease in ribulose-1,5-bisphosphate carboxylase (Rubisco) at the gene, protein and activity levels (Walters and Ayres, 1984; Ranty *et al.*, 1987; Mouly *et al.*, 1988; Roby *et al.*, 1988a; Chapter 7, section 7.3.3). Changes in the amounts of both the small and the large subunits were readily observable after gel electrophoresis (Figure 5.1). Disappearance of the two subunits and their transcripts were inversely correlated with the appearance of defence proteins, occurring earlier and to a greater extent in susceptible than in tolerant cultivars. Indeed, it has been reported that ethylene and jasmonic acid, two possible signals in the transduction cascade following host–pathogen recognition, lead to decreases in Rubisco gene expression (Weidhase *et al.*, 1987; Mueller-Uri *et al.*, 1988). This suggests that metabolic alterations and defence induction are not separate phenomena but instead share common determinants. Finally, it must be emphasized that, in contrast to the early and transient expression of signalling molecules, changes in defence proteins and metabolic markers are delayed, but long-lasting events.

5.5 Conclusion

This chapter has been concerned mainly with changes in proteins whose functions are already known. Most of these proteins are also induced under particular developmental and environmental conditions, which suggests that they might have multiple functions. Some of them might have a basic developmental role but additional functions might have developed in the course of evolution. It is noteworthy that, upon infection, they are efficiently induced in those cultivars where the presence of resistance genes leads to cultivar-specific resistance. The phenomenon might depend on molecular events not always requiring protein synthesis. An overview and intra- and intercellular signalling pathways resulting in the induction of defence proteins is proposed in Figure 5.3. Future work will look for such specific changes.

References

Benhamou, N., Mazau, D. and Esquerré-Tugayé M.T. (1990) Immunogold localization of hydroxyproline-rich glycoproteins in necrotic tissue of *Nicotiana tabacum* L. cv. Xanthi-nc infected by tobacco mosaic virus. *Physiol. Mol. Plant Pathol.* **36**, 129–145.

Benhamou, N., Mazau, D., Grenier, J. and Esquerré-Tugayé, M.T. (1991a) Time-course study of the accumulation of hydroxyproline-rich glycoproteins in root cells of susceptible and resistant tomato plants infected by *Fusarium oxysporum* f.sp. *radicis-lycopersici. Planta* **184**, 196–208.

Benhamou, N., Lafontaine, P.J., Mazau, D. and Esquerré-Tugayé, M.T. (1991b) Differential accumulation of hydroxyproline-rich glycoproteins in bean root nodule cells infected with a wild-type strain or a C4-dicarboxylic acid mutant of *Rhizobium leguminosarum* pv. *phaseoli. Planta* **184**, 457–467.

Benhamou, N., Lafitte, C., Barthe, J.P. and Esquerré-Tugayé, M.T. (1991c) Cell surface interactions between bean leaf cells and *Colletotrichum lindemuthianum*: cytochemical aspects of pectin breakdown and fungal endopolygalacturonase accumulation. *Plant Physiol.* **97**, 234–244.

Bonfante-Fasolo, P., Tamagnone, L., Peretto, R., Esquerré-Tugayé, M.T., Mazau, D., Mosiniak, M. and Vian, B. (1991) Immunocytochemical location of hydroxyproline-rich glycoproteins at the interface zone between mycorrhizal fungus and host plants. *Protoplasma* **165**, 127–138.

Broglie, K.E., Gaynor, J.J. and Broglie R.M. (1986) Ethylene-regulated gene expression: molecular cloning of the genes encoding an endochitinase from *Phaseolus vulgaris. Proc. Natl. Acad. Sci. USA* **83**, 6820–6824.

Cervone, F., Hahn, M.G., De Lorenzo, G., Darvill, A. and Albersheim, P. (1989) Host-pathogen interactions. XXXIII. A plant protein converts a fungal pathogenesis factor into an elicitor of plant defense responses. *Plant Physiol.* **90**, 542–548.

Chappell, J., Hahlbrock, K. and Boller, T. (1984) Rapid induction of ethylene biosynthesis in cultured parsley cells by fungal elicitor and its relationship to the induction of phenylalanine ammonia-lyase. *Planta* **161**, 475–480.

Collinge, D.B. and Slusarenko, A.J. (1987) Plant gene expression in response to pathogens. *Plant Mol. Biol.* **9**, 339–410.

Corbin, C.R., Sauer, N. and Lamb, C.J. (1987) Differential regulation of a hydroxyproline-rich glycoprotein gene family in wounded and infected plants. *Mol. Cell Biol.* **7**, 4337–4344.

Daugrois, J.-H., Lafitte, C., Barthe, J.P. and Touzé, A. (1990) Induction of a β-1,3-glucanase and chitinase activity in compatible and incompatible interactions between *Colletotrichum lindemuthianum* and bean cultivars. *J. Phytopathol.* **130**, 225–234.

Daugrois, J.-H., Lafitte, C., Barthe, J.P., Faucher, C., Touzé, A. and Esquerré-Tugayé, M.T. (1992) Purification and characterization of two basic β-1,3-glucanases induced in *Colletotrichum lindemuthianum* infected bean seedlings. *Arch. Biochem. Biophys.* **292**, 468–474.

Esquerré-Tugayé, M.T. (1973) Influence d'une maladie parasitaire sur la teneur en hydroxyproline des parois cellulaires d'épicotyles et pétioles de plantes de Melon. *C.R. Acad. Sci., Paris* **276**, 525–528.

Esquerré-Tugayé, M.T. and Mazau, D. (1974) Effect of a fungal disease on extensin, the plant cell wall glycoprotein. *J. Exp. Bot.* **25**, 509–513.

Esquerré-Tugayé, M.T., Lafitte, C., Mazau, D., Toppan, A. and Touzé, A. (1979) Cell surfaces in plant-microorganism interactions. II. Evidence for the accumulation of hydroxyproline-rich glycoproteins in the cell wall of diseased plants as a defense mechanism. *Plant Physiol.* **64**, 320–326.

Farmer, E.E. and Ryan, C.A. (1990) Interplant communication: airborne methyl jasmonate induces synthesis of proteinase inhibitors in plant leaves. *Proc. Natl. Acad. Sci. USA* **87**, 7713–7716.

Graham, J.S., Hall, G., Pearce, G. and Ryan, C.A. (1986) Regulation of synthesis of proteinase inhibitors I and II mRNAs in leaves of wounded tomato plants. *Planta* **169**, 399–405.

Heath, M.C. (1986) Fundamental questions related to plant-fungal interactions: can recombinant DNA technology provide the answers? In: *Biology and Molecular Biology of Plant-Pathogen Interactions* (ed. J.A. Bailey). Springer Verlag, Berlin Heidelberg, pp. 15–27.

Johnson, R., Narvaez, J., An, G. and Ryan, C. (1989) Expression of proteinase inhibitors I and II in transgenic tobacco plants: effects on natural defense against *Manduca sexta* larvae. *Proc. Natl. Acad. Sci. USA* **86**, 9871–9875.

Katou, K., Tomiyama, K. and Okamoto, H. (1982) Effects of hyphal wall components of *Phytophthora infestans* on membrane potential of potato tuber cells. *Physiol. Plant Pathol.* **21**, 311–317.

Kauffmann, S., Legrand, M., Geoffroy, P. and Fritig, B. (1987) Biological function of "pathogenesis-related" proteins: four PR proteins of tobacco have 1,3-β-glucanase activity. *EMBO J.* **6**, 3209–3212.

Lafitte, C., Barthe, J.P., Montillet, J.L. and Touzé, A. (1984) Glycoprotein inhibitors of *Colletotrichum lindemuthianum* endopolygalacturonase in near isogenic lines of *Phaseolus vulgaris* resistant and susceptible to anthracnose. *Physiol. Plant Pathol.* **25**, 39–53.

Mayer, M.G. and Ziegler, E. (1988) An elicitor from *Phytophthora megasperma* f.sp. *glycinea* influences the membrane potential of soybean cotyledonary cells. *Physiol. Mol. Plant Pathol.* **33**, 397–407.

Mazau, D. and Esquerré-Tugayé, M.T. (1986) Hydroxyproline-rich glycoprotein accumulation in the cell walls of plants infected by various pathogens. *Physiol. Mol. Plant Pathol.* **29**, 147–157.

Mazau, D., Rumeau, D. and Esquerré-Tugayé, M.T. (1988) Two different families of hydroxyproline-rich glycoproteins in melon callus: biochemical and immunochemical studies. *Plant Physiol.* **86**, 540–546.

Miyazaki, J.H. and Yang, S.F. (1987) The methionine salvage pathway in relation to ethylene and polyamine biosynthesis. *Physiol. Plant.* **69**, 366–370.

Mouly, A., Roby, D. and Esquerré-Tugayé, M.T. (1988) Ribulose-1,5- bisphosphate carboxylase-oxygenase small subunit transcripts as a susceptibility reflecting molecular marker in sunflower infected with *Sclerotinia sclerotiorum*. *Plant Sci.* **56**, 219–225.

Mueller-Uri, F., Parthier, B. and Nover, L. (1988) Jasmonate-induced alteration of gene expression in barley leaf segments analyzed by *in vivo* and *in vitro* protein synthesis. *Planta* **176**, 241–247.

Neale, A.D., Wahleithner, J.A., Lund, M., Bonnett, H.T., Kelly, A., Meeks-Wagner, D.R., Peacock, W.J. and Dennis, E.S. (1990) Chitinase, β-1,3-glucanase, osmotin, and extensin are expressed in tomato explants during flower formation. *Plant Cell* **2**, 673–684.

O'Connell, R.J., Brown, I.R., Mansfield, J.W., Bailey, J.A., Mazau, D., Rumeau, D. and Esquerré-Tugayé M.T. (1990) Immunocytochemical localization of hydroxyproline-rich glycoproteins accumulating in melon and bean at sites of resistance to bacteria and fungi. *Mol. Plant Microb. Interact.* **3**, 33–40.

Ohta, H., Shida, K., Peng, Y.L., Furasawa, I., Shishiyama, J., Aibara, S. and Morita, Y. (1991) A lipoxygenase pathway is activated in rice after infection with the rice blast fungus *Magnaporthe grisea*. *Plant Physiol.* **97**, 94–98.

Pélissier, B., Thibaud, J.B., Grignon, C. and Esquerré-Tugayé, M.T. (1986) Cell surface in plant-microorganism interactions. VII. Elicitor preparations from two fungal pathogens depolarize plant membranes. *Plant Sci.* **46**, 103–109.

Pena-Cortès, H., Sanchez-Serrano, J.J., Mertens, R., Willmitzer, L. and Prat, S. (1989) Abscisic acid is involved in the wound-induced expression of the proteinase inhibitor II gene in potato and tomato. *Proc. Natl. Acad. Sci. USA* **86**, 9851–9855.

Ranty, B., Roby, D., Cavalié, G. and Esquerré-Tugayé, M.T. (1987) Ribulose 1–5 bisphosphate carboxylase/oxygenase expression in melon plants infected with *Colletotrichum lagenarium*: activity, content, and rate of synthesis. *Planta* **170**, 386–391.

Rickauer, M., Fournier, J., Pouénat, M.L., Berthalon, E., Bottin, A. and Esquerré-Tugayé, M.T. (1990) Early changes in ethylene synthesis and lipoxygenase activity during defense induction in tobacco cells. *Plant Physiol. Biochem.* **28** (5), 647–653.

Rickauer, M., Bottin, A. and Esquerré-Tugayé, M.T. (1992) Regulation of proteinase inhibitor induction in tobacco cells by fungal elicitors, hormonal factors and methyl jasmonate. *Plant Physiol. Biochem.* (in press).

Roby, D. and Esquerré-Tugayé, M.T. (1986) Cell surfaces in plant-microorganism interactions. VI. Elicitors of ethylene from *Colletotrichum lagenarium* trigger chitinase activity in melon plants. *Plant Physiol.* **81**, 228–233.

Roby, D. (1987) *Réactions de défense et expression génique de plantes de melon en réponse á une infection fongique*. Thèse Doctorat d'Etat Sciences Naturelles, Université Toulouse.

Roby, D., Toppan, A. and Esquerré-Tugayé, M.T. (1987) Cell surfaces in plant-microorganism interactions. VIII. Increased inhibitor proteinase activity in melon plants in response to infection by *Colletotrichum lagenarium* or to treatment with an elicitor fraction from this fungus. *Physiol. Mol. Plant Pathol.* **30**, 453–460.

Roby, D., Marco, Y. and Esquerré-Tugayé, M.T. (1988a) Ribulose 1,5-bisphosphate carboxylase/oxygenase in melon plants infected with *Colletotrichum lagenarium*. Activity, content and rate of synthesis of mRNAs coding for the large and small subunits. *Physiol. Mol. Plant Pathol.* **32**, 411–424.

Roby, D., Toppan, A. and Esquerré-Tugayé, M.T. (1988b) Systemic induction of chitinase activity and resistance in melon plants upon fungal infection or elicitor treatment. *Physiol. Mol. Plant Pathol.* **33**, 409–417.

Rumeau, D., Mazau, D. and Esquerré-Tugayé, M.T. (1987) Cytosine-rich RNAs from infected plants and their *in vitro* translation products. *Physiol. Mol. Plant Pathol.* **31**, 305–313.

Rumeau, D., Mazau, D., Panabières, F., Delseny, M. and Esquerré-Tugayé, M.T. (1988)

Accumulation of hydroxyproline-rich glycoprotein mRNAs in infected or ethylene treated melon plants. *Physiol. Mol. Plant Pathol.* **33**, 419–428.

Ryan, C.A. (1990) Protease inhibitors in plants: genes for improving defenses against insects and pathogens. *Ann. Rev. Phytopathol.* **28**, 425–449.

Showalter, A.M., Bell, J.N., Cramer, C.L., Bailey, J.A., Varner, J.E. and Lamb, C.J. (1985) Accumulation of hydroxyproline-rich glycoprotein mRNAs in response to fungal elicitor and infection. *Proc. Natl. Acad. Sci. USA* **82**, 6551–6655.

Toppan, A. and Esquerré-Tugayé, M.T. (1984) Cell surfaces in plant-microorganism interactions. IV. Fungal glycopeptides which elicit the synthesis of ethylene in plants. *Plant Physiol.* **75**, 1133–1138

Walters, D.R. and Ayres, P.G. (1984) Ribulose bisphosphate carboxylase protein and enzymes of CO_2 assimilation in barley infected by powdery mildew (*Erysiphe graminis hordei*). *Phytopath. Z.* **109**, 208–218.

Weidhase, R.A., Lehmann, J., Kramell, H., Sembdner, G. and Parthier, B. (1987) Degradation of ribulose-1,5-bisphosphate carboxylase and chlorophyll in senescing barley leaf segments triggered by jasmonic acid methylester, and counteraction by cytokinin. *Physiol. Plant.* **69**, 161–166.

Ye, Z.H., Song, Y.R., Marcus, A. and Varner, J.E. (1991) Comparative localization of three classes of cell wall proteins. *The Plant Journal* **1**, 175–183.

6

Resistance and defence: the role of secondary plant substances

P.J. Edwards

6.1 Secondary plant substances and the defence paradigm

For the past forty years the paradigm within which biologists have studied secondary plant substances has been one which assumes that their primary role is as defences against herbivores and pathogens. An important step in establishing this paradigm was the paper by Fraenkel (1959) in which he noted, not only the widespread toxicity of secondary metabolites to insects, but also the dependence by adapted insects upon particular metabolites for locating their host plants and as a cue for feeding. For example, the diamond-back moth, *Plutella maculipennis*, feeds exclusively on Cruciferae and a few species of other families, all of which contain mustard oil glucosides, and can only be induced on other plants if the glucoside is added. Fraenkel concluded that these substances have evolved as defences against plant-feeding insects, though insects which overcome them have evolved to use them for their own purposes. Thus, an 'adapted' insect may use the secondary substances of the host species to locate a food plant, to defend itself against attack, or as a basis for the production of pheromones (Harborne, 1988).

Subsequently, Ehrlich and Raven (1964) proposed a theory of co-evolution between plants and the insects which feed on them, which provided a mechanism to explain the patterns of association between plants and insects. They suggested that plants have, through occasional mutation and recombination, produced novel compounds not directly related to basic metabolism. Some of these compounds were toxic to phytophagous insects, and the plants thus defended entered a new 'adaptive zone' in which, released from the adverse effects of herbivory, they were able to diversify. However, insects which evolved to overcome the physiological obstacles presented by secondary plant substances were themselves free to diversify, largely in the absence of competition from other plant-feeding insects. Thus co-evolution was seen as a process of adaptation and counteradaptation by both plant and insect.

The great attraction of the theory of co-evolution was that it explained many hitherto puzzling phenomena, such as the great diversity of secondary plant substances, and the fact that most plant-feeding insects use relatively few, and often related, plant species. The widespread acceptance of the theory had profound consequences for the subsequent direction of research in the field of insect–plant interactions. Indeed so pervasive has been its influence that the study of the role of secondary substances is often referred to as 'plant defence theory', and the terms 'defensive compound' and 'secondary plant substance' are used more or less interchangeably (Edwards, 1989). Proceeding from an assumption of the primary importance of insect herbivores as agents of natural selection, ecologists have devoted considerable efforts towards developing a general theory of plant defence which can explain patterns of variation in the types and quantities of secondary plant compounds (and physical defences) in terms of the need for protection against insect herbivores.

Two theories have been particularly influential. According to the apparency hypothesis (Feeny, 1976), the risk of discovery by herbivores is the major factor influencing the extent of commitment by a plant to defence, and also the type of defence. Apparent plants are those with a high risk of discovery (for example, because they are large or otherwise conspicuous); such plants have a high investment in quantitative defences such as tannins or silica, which, without being highly toxic, are broadly effective against a wide range of organisms. In contrast, unapparent plants are more likely to escape grazing, and therefore employ much smaller quantities of toxins, which although highly efficient against non-adapted enemies, are in an evolutionary sense easy to overcome.

An alternative hypothesis to explain the differences between plant species in their defensive characteristics is that these differences are related to the intrinsic rate of growth of plants (Coley *et al.*, 1985). According to this hypothesis, slow growing plants of resource-poor environments tend to have high levels of immobile defences such as tannins and lignins. Such materials have high initial construction costs but do not have continued metabolic costs of turnover, so that their cost is independent of leaf lifetime. In contrast, faster growing plants characteristic of resource-rich environments are characterized by smaller concentrations of mobile compounds of low molecular weight, such as alkaloids and cardiac glycosides, which since they have rapid turnover rates must be continually synthesized.

Fundamental to both hypotheses is the notion that secondary plant compounds are costly to the plant (Coley, 1986; Gulmon and Mooney, 1986; Fagerstrom *et al.*, 1989), and that the allocation of the resources to produce them has been determined through natural selection, chiefly by insect herbivores. Thus, as Parker describes in Chapter 12, both hypotheses are in an adaptationist tradition which 'advocates that resistance can best be undestood by a cost-benefit analysis where the benefits of limiting parasite attack are balanced by various types of fitness costs associated with resistance'.

This chapter presents the case against the defence paradigm. It argues that the evolution and diversification of secondary plant metabolism has not been primarily a response to biotic factors such as insect herbivores and plant pathogens. However, defensive systems have evolved, against both insects and micro-organisms, which take advantage of metabolic pathways leading to the production of toxic secondary substances.

6.2 Critique of defence theory

6.2.1 *Co-evolution between plants and phytophagous insects*

In recent years Ehrlich and Raven's (1964) hypothesis of co-evolution between plants and insects, and the conclusions drawn about the defensive role of secondary substances, have both come under close critical scrutiny. The capacity of insect herbivores to exert a significant influence on the course of plant evolution remains the subject of continuing controversy. Thus, Strong *et al.* (1984) argued that the majority of insect species probably have little impact on the performance of their food plants, because their populations are limited by the vagaries and harshness of the environment or by the impact of natural enemies. Similarly, Jermy (1984) maintained that the evolutionary interactions between plants and phytophagous insects are asymmetric. While plants, through their biochemical and structural diversity, provide a profusion of niches for the evolutionary radiation of insects, insects do not affect plant evolution, or at most may cause anagenic changes in the plant. In Jermy's view, insects could not have been the main cause of the appearance and maintenance of allelo-chemicals in plants, though these substances have profound implications for insects since they determine the biochemical profile by which they distinguish between host and non-host plants.

There is some evidence, however, that the acquisition of more effective defences has been important in the success of particular taxa. For example, Berenbaum (1983) demonstrated that groups of genera in the Umbelliferae, characterized by coumarins of increasing advancement, showed increasing mean numbers of species per genus. She concluded that the evolution of greater toxicity in this family has been associated with more rapid speciation, as expected from co-evolution theory. Recently Farrell *et al.* (1991) set out to test the hypothesis that the evolution of a new defensive type may lead to evolutionary success, by comparing the diversity of 16 plant groups possessing latex or resin canals (assumed to be defensive against insects and micro-organisms) with that in sister groups lacking such structures. For example, within the order Gentianales, the canal-bearing families Apocynaceae and Asclepiadaceae both have around 2000 species, while the Gentianaceae and Loganiaceae, in which canals are absent, have 900 and 600 species, respectively. This analysis showed that, consistent with the hypothesis of there being an evolutionary advantage from the acquisition of a defence, canal-bearing lineages had considerably higher diversity than their sister groups.

As in other evolutionary arguments, the debate about insect–plant co-evolution has been hampered by the nature of the evidence; we must make our inferences about past evolutionary processes from evidence provided by the ecological interactions between present day plants and insects, about whose evolutionary history we know very little. However, new methods of establishing the evolutionary relationships between taxa are being developed which offer important new approaches to studying the evolution of insect–plant relationships (Mitter *et al.*, 1991). The determination of phylogenetic relationships of both plant and insect taxa, using cladistic analysis of molecular data, offers the opportunity to examine to what extent the phylogenies of the two groups are congruent. Although there is some evidence that the relative ages of origin of the major clades of phytophagous insects correspond with those of their major host clades, the strict congruence of phylogenies (which would occur in the case of true co-evolution) appears to be rare. It is clear from such studies that host shifts have occurred frequently for many groups of insects, though the shifts have in general been constrained to new hosts which are chemically similar to the old hosts. Such evidence tends to support the conclusion that many insect–plant associations are relatively recent, and have come about through the opportunistic colonization by insects of chemically suitable hosts (Strong *et al.*, 1984).

6.2.2 *Role of secondary substances*

Another area of continuing debate concerns the role of secondary substances. The issue is not so much whether secondary substances act as defences against insects – there is a wealth of evidence to show that they do – but whether their primary role is defensive, and if so, whether insects represent the principal selective pressure. Indeed, a valid criticism of plant defence theory is that it has failed to give due attention to other organisms, and in particular pathogens. This situation has arisen because of the distinct, and to a large extent independent, approaches to the study of plant–herbivore and plant–pathogen interactions; whereas those interested in herbivory have tended to focus upon the mechanisms of resistance (including the role of secondary compounds), plant–pathogen work has been more concerned with the genetic basis of resistance and virulence. Thus, 'in plant–herbivore studies, much work has been done on how resistance factors affect the insect's attraction to the plant, its ability to digest plant tissue, and the diet's effect on insect growth and reproduction. Similar detail on the effects of host plants on pathogen biology is lacking.' (Marquis and Alexander, 1992).

Despite this relative neglect of the role of secondary chemistry in plant–pathogen studies, a strong case can be made for regarding secondary compounds as of considerable importance as defences against fungi. Firstly, in an evolutionary timescale, attack from pathogens has been a threat to plants for a much longer period (Pirozynski, 1976) than the threat from insects (or even from more primitive arthropods; Kevan *et al.*, 1976), and the evolution of many secondary compounds certainly predates the appearance of the insects. Secondly, there is accumulating evidence that fungi in natural communities do adversely affect the population dynamics of many plant species. Finally, many of the attributes of plants commonly regarded as anti-herbivore defences also confer resistance against pathogens. Thus, there is no *a priori* reason to regard insects as the principal selective factor for resistance in all circumstances.

An example of the difficulty in distinguishing between possible defence against herbivores and pathogens is provided by tannins. Although tannins are widely regarded as defensive compounds against insects, Bernays (1981) concluded that it is difficult to generalize about their ecological and evolutionary significance. While those species of insects which do not normally feed on tannins may be adversely affected, insects which feed habitually on trees and shrubs are often stimulated to feed by hydrolysable tannins. Thus, in bioassay experiments the palatability and digestibility of plant food to the tree locust, *Anacridium melanorhodon*, was increased if tannic acid was added to it (Bernays, 1981). Clearly this group of compounds cannot be regarded as quantitative defences against insects of the kind proposed in the apparency hypothesis (Feeny, 1976), which always adversely affect growth in a dosage-dependent manner because of their anti-digestive action. However, there is correlative evidence concerning tannin levels and the relative vulnerability of plants to certain pathogens (Scalbert, 1991; Zucker, 1983). Furthermore, the most intense tannin deposition in leaves tends to occur in the outer epidermal cell wall of leaves, where it could be most effective in preventing entry of fungal pathogens. For these reasons, Bernays *et al.* (1989) suggest, 'as a working hypothesis', that tannins are more important as defences against pathogens than against insect herbivores.

A strong case can also be made for regarding certain secondary compounds as defences against vertebrate herbivores (Grubb, 1992). For example, one has only to see how certain toxic or unpalatable species (e.g. *Senecio jacobaea*, *Hyoscyamus niger*, *Euphorbia amygdaloides*) are avoided by livestock, to appreciate that secondary substances may provide extremely effective defence. Furthermore, the activity of some of these substances (for example, the isoflavonoids of many leguminous species, which mimic mammalian reproductive hormones) appears to be specifically against mammalian herbivores (Heltmann, 1975; Shutt, 1976).

The weakness of this adaptationist approach is that it attempts to assign specific adaptive functions to particular metabolic products. It would be more realistic to recognize that the selective pressures faced by plants come from many directions: from vertebrate and invertebrate herbivores, from pathogens, from other plants, and from the abiotic environment. The evolution of a biosynthetic pathway leading to the accumulation of a highly active compound is likely to affect to some degree a plant's interactions with all these components of its environment. For example, highly toxic compounds such as the hydroxamic acids found in certain grasses have been shown to be active against a very wide range of organisms, including bacteria, fungi, nematodes and insects (Niemeyer and Perez, 1987). Synthesis of cucurbitacins in *Cucurbita moschata* is stimulated by exposure to UV radiation, and these compounds are effective as filters of UV radiation; however these substances are extremely toxic to insects, and synthesis may also be stimulated by wounding (Tallamy and Krischik, 1989). Indeed, the universal occurrence of flavonoids in land plants may be due, at least in part, to their high absorbtivity of ultraviolet radiation and their role as antioxidants (Swain, 1975). Thus, very rarely can we assign a specific and unique function to a secondary metabolic product, and we should be wary of attempts to do so.

Finally, in searching for the adaptive significance of secondary metabolites, we should not ignore the possibility that the production of certain metabolites is the result of physiological constraints upon primary metabolic activity. Thus it has been suggested that the high concentrations of carbon-based secondary substances commonly found in plants of nutrient-poor habitats are not primarily defensive, but rather a metabolic consequence of nutrient deficiency. According to the nutrient stress hypothesis (Tuomi *et al.*, 1984, 1988), carbon is allocated to new growth when the supply of mineral nutrients is adequate for new cells, but surplus carbon may be converted into allelo-chemicals. Similar physiological constraints may also explain the generally higher level of phenolic compounds in sun leaves relative to shade leaves. Since changes in the secondary chemistry of plants are often associated with other changes which may affect insects, interpreting the effect of secondary substances upon insect performance is often difficult. For example, Feeny (1970) linked the reduced feeding by winter moth caterpillars on oak leaves with the increasing level of phenolics in the leaves. However, an equally plausible interpretation is that decreased insect predation on oak in early summer is related to changes in cell structure, texture, and toughness of the leaf, in the development of which the formation of phenolic polymers plays an intimate role (Haslam, 1988).

In conclusion, nearly 30 years after the publication of Ehrlich and Raven's theory of co-evolution (1964), our views of the role of insects in the evolution of plant defences remain unclear and confused. There may be a few examples of genuine co-evolution, but, to many biologists, these examples 'look increasingly like small vortices in the mainstream of evolution' (Strong *et al.*, 1984). If insects do affect the course of evolution, it is often likely to be through diffuse co-evolution, in which plants evolve to cope with attacks of all kinds from a wide range of different organisms, including insects, vertebrates and micro-organisms (Fox, 1981). Fraenkel's (1959) bold generalization that the 'raison d'être' of plant secondary metabolites is as defences against insects appears increasingly insecure.

6.3 Metabolic interpretation of the diversity of secondary plant compounds

Despite the difficulties with defence theory, alternative explanations for the role of secondary metabolites, for example that they represent metabolic waste products, have received little support. The objections to such explanations are well illustrated in an exchange between Muller (1969) and Janzen (1969). Muller argued that secondary metabolites are primarily metabolic wastes, though recognizing that other organisms, such as herbivores and competing plants, may act selectively upon them. 'Thus the toxicity to animals of these metabolic wastes, no matter how important eventually, is subsequent and secondary to their elimination from protoplasm.' Janzen replied that 'it is hard to understand how many compounds (for example, alkaloids, free amino acids, saponins, glycosides and so forth), virtually unique to plant metabolic systems, can be considered waste products, when animal metabolic systems do quite well with very few kinds of waste products . . . What other selective force in the environment besides herbivores (*sensu lato*) and competing plants has the diversity of quality, yet specific persistence of environmental challenge to lead to essentially a unique combination and array of secondary compounds for each species of plant?'

Recently Gottlieb (1990) has proposed a metabolic basis for the diversification of secondary plant metabolites which, it is believed, for the first time seriously challenges

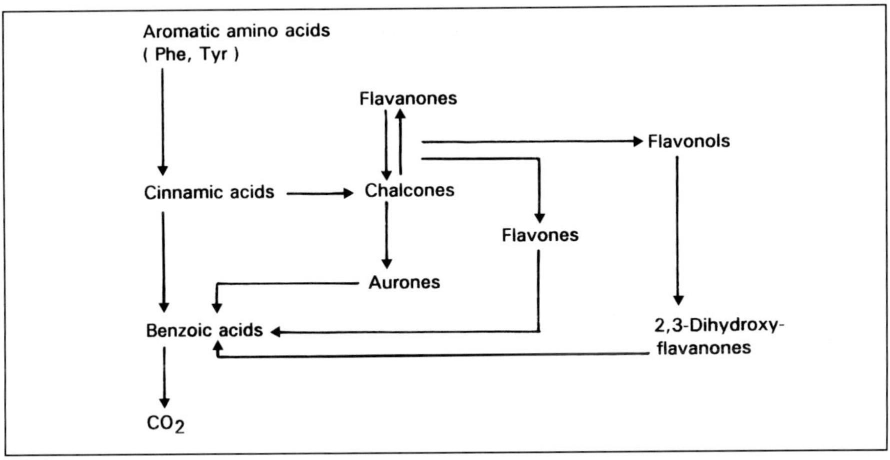

Figure 6.1. *Derivation of various biosynthetic groups of secondary metabolites from cinnamic acids. Reproduced from Gottlieb (1990) with permission from Pergamon Press Ltd.*

the defence paradigm as the 'raison d'être' of these substances. Gottlieb argues that the diversity of secondary metabolites (which he prefers to call special metabolites) amongst plants can largely be explained by the consequences of internal driving forces leading to the evolution of more efficient metabolic systems. At the core of this hypothesis is the notion, put forward by Bu'Lock (1980) in the context of micro-organisms, that 'secondary biosynthesis serves to maintain basic metabolism in circumstances when its products (high energy phosphates, reduced co-enzymes, and substrates for synthesis) cannot – through nutritional imbalances – be used for cell replication, and to do this in a way which can be safely turned off when environmental conditions improve . . .'

One way in which basic metabolism could be maintained by secondary metabolism is through controlling the rate of turnover of metabolic products under conditions which are not conducive to biosynthesis. Primary metabolic pathways can bring about the complete turnover of organic materials and, in the absence of secondary metabolism, metabolic intermediates may be rapidly broken down to carbon dioxide. However, if metabolites are shunted into secondary metabolic pathways, the rate of decomposition of intermediates may be considerably reduced. For example, cinnamic acids, intermediates of the primary shikimate pathway, are degraded by very few steps to benzoic acids and thus to carbon dioxide, unless obliged to pass through a long series of intermediates after protective condensation with acetic acid (Figure 6.1).

Consistent with this interpretation of the role of secondary metabolism is the fact that secondary metabolic pathways are usually closely linked with primary metabolism. Indeed, part of the explanation for the diversification of secondary metabolism in terrestrial plants may lie in the changes which have occurred in the relative importance of the various primary metabolic pathways. In woody plants, the shikimate pathway, which produces lignin, is of major importance, and the predominant

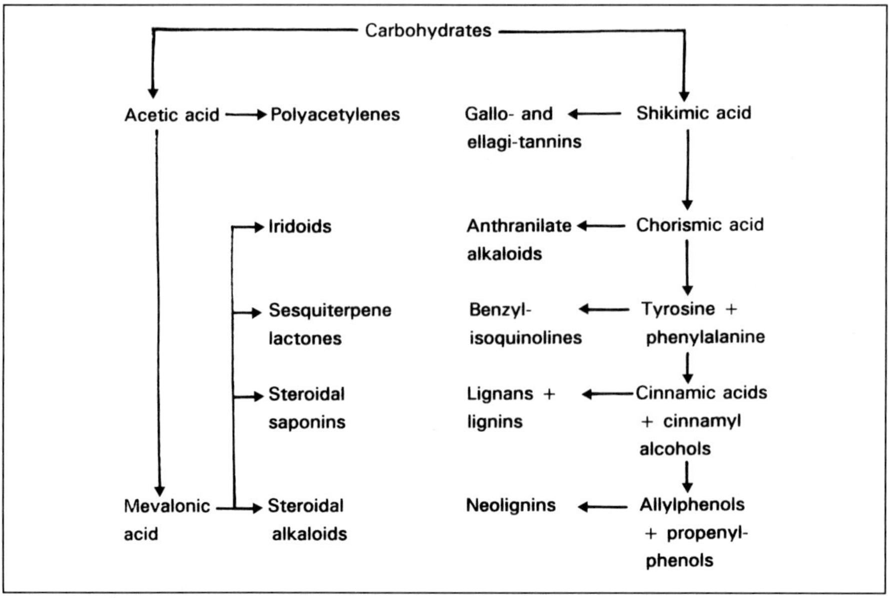

Figure 6.2. *Biosynthetic pathways of major groups of secondary metabolites, showing the close link between secondary and primary metabolism. Reproduced from Gottlieb (1989) with permission from Pergamon Press Ltd.*

secondary metabolites of woody plants are themselves products of that pathway (for example condensed tannins and benzylisoquinoline alkaloids; Figure 6.2). In contrast in herbaceous plants, in which the shikimate pathway is less dominant, we find a tendency towards secondary metabolites which are derived, either from the earlier steps of the shikimate pathway, or from the acetate–mevalonate pathway (e.g. iridoid alkaloids and sesquiterpene lactones).

Another factor which may have led to diversification of secondary metabolism is the influence of atmospheric oxygen upon the efficiency of metabolism. There is a remarkably close correlation amongst many biosynthetic groups of secondary metabolites between the degree of skeletal specialization of compounds (i.e. the degree to which they have been modified compared with the precursor of the group) and their level of oxygenation (Gottlieb, 1989). There is also often a clear correlation between the level of oxygenation in secondary metabolites and the evolutionary advancement of the plant taxon in which they occur. For example, within the Magnolidae there is a strong positive correlation between the mean level of oxygenation of benzylisoquinoline alkaloids and the index of advancement (Sporne, 1980) of the families. These trends suggest that there has been selection for progressively higher levels of oxygenation of secondary metabolites during the course of plant evolution.

Gottlieb argues that the trend towards higher levels of oxygenation in secondary compounds is also consistent with the hypothesis that the role of secondary metabolism is to maintain primary metabolism, since one aspect of this role is that secondary metabolites may act as antioxidants, preventing the adventitious oxidation of primary metabolites. The driving force behind progressively higher levels of oxidation of secondary metabolites may lie in the smaller amounts of energy required to convert them to starting materials, thus making them more effective as antioxidants, and providing for greater economy in recovering materials during senescence. However, highly oxidized compounds tend to be unstable, and thus their turnover rate will be higher unless they are protected from oxidation in some way. The more oxidized secondary compounds often exhibit biochemical protective devices, such as *o*-methylation of phenolic groups, which diminish the rate of turnover. The variety of such protective devices itself represents a source of structural variants, and hence of secondary metabolite diversity.

6.4 Secondary roles for secondary metabolites

If the primary role of secondary metabolism is the maintenance of basic metabolism, as Bu'Lock (1980) argues, then it is the *processes* of biosynthesis that are beneficial, rather than the production of any particular metabolite. The diversification of secondary products has come about through an adaptation of these biosynthetic processes to changing patterns of primary metabolism, and to changing environmental conditions, notably the increase in atmospheric oxygen. However this concept of the selective value of secondary metabolism *per se* in no way excludes the possibility that plants may derive additional selective advantage from the possession of the metabolites thus produced. Given the enormous diversity of biosynthetic groups of compounds produced by plants, it would be extraordinary if some did not prove to be of selective value in some aspect of the plant's metabolism or interactions with its environment. During the course of plant evolution, it is probable that the elaboration of basic metabolism has occurred through the selection of 'secondary' metabolites which first appeared by the processes of diversification such as those outlined above. Examples of such metabolic processes from the early land plants include the pathways leading to the production of cutin, suberin and lignin. Amongst the additional selective advantages which a plant may receive by the possession of a particular biochemical pathway is the opportunity to use its products for defence against organisms which attack it or compete with it. Given that secondary metabolism is a general property of all plants, the problem arises of how we might distinguish a *defensive system* which takes advantage of a secondary metabolic pathway from the mere presence of the biosynthetic product. While it is always dangerous to ascribe an adaptive function to particular structures, it is probable that any defensive system would involve at least one, and probably several, of the following features:

(1) secondary metabolites concentrated in the most important or vulnerable plant tissues,
(2) synthesis greatest at the most vulnerable stages of growth,
(3) production of special storage structures,
(4) production of transport structures,
(5) enhanced synthesis in response to infection or attack,
(6) production and transport of elicitors in response to infection or attack,
(7) selection of the most toxic compounds in biosynthetic pathways.

Table 6.1. *Contrasts in the exploitation of plants by insects and fungi and in the effects that they have upon their host*

	Insects	Fungi
Mobile foraging	Yes	No
Host location	Active, sensory	Random dispersal of spores
Host recognition	Sensory	Chemical
Penetration of tissues	Mechanical	Enzymic/ mechanical
Mineral nutrient requirements	High, especially for nitrogen	Can be very low
Life span	Short	Short to very long
Distribution of damage resulting from attack	Local or dispersed	Local or extensive
Effect of damage	Cumulative effects of local damage	May have systemic effects

6.5 Contrasting defensive systems against insects and micro-organisms

This section speculates upon the circumstances favouring evolution of defensive systems against insects and fungi, and compares the likely characteristics of such systems.

6.5.1 *Factors affecting the exploitation of plants*

There is a broad correspondence in the plant resistance factors effective against insects and pathogens. Thus, secondary substances such as gallic acid, chlorogenic acid, caffeic acid, mustard oil glycosides, alkaloidal saponins and cyanogenic glycosides have been shown to confer resistance against fungi as well as against insects (Mansfield, 1983). Similarly, characteristics associated with tough leaves, such as a thick cuticle, the presence of silica, and a high proportion of lignin, are resistance factors to both insects and fungi. Such similarities in resistance factors are not surprising. Both herbivores and micro-organisms face similar physical problems of penetrating their host's tissues, and similar chemical problems associated with the toxic action of secondary metabolites. Furthermore, in as much as such obstacles represent evolved defences, they may be the product of diffuse co-evolution, and thus an evolutionary response to attack from insects and micro-organisms alike.

However, despite broad similarities in plant resistance factors against micro-organisms and insects, there are also important differences, which arise from the fundamental differences in their physiology and ecology, and thus in the nature of the threat they present to plants (Table 6.1).

The characteristics of insects which make them pests of plants include their short lifecycle and high intrinsic rates of population increase. However, to achieve high rates of population growth, insects must obtain food of adequate quality, and in this respect most plant food is less than optimal, particularly in its nitrogen content (Mattson, 1980). Whereas the nitrogen contents of plant tissues can range from 0.03 to 7.0 % of dry weight, animals contain 8 to 14 % of nitrogen, and so plant feeders must concentrate it from their foodstuffs. Herbivores differ in their capacity to feed on plant material of low nitrogen content, and particular species exhibit clearly defined minimum requirements. For example, the *Paropsis* leaf beetle on Eucalyptus foliage appears to need at least 1.0 % nitrogen in its diet, while some grasshoppers need about 3 %.

Partly because of the importance of food quality, insects often exhibit highly elaborate food selection behaviour in which both visual and chemosensory cues play a part. Having alighted on a host plant, the decision of whether to feed (or to lay eggs) is based upon a complex series of sensory inputs in which the secondary chemical profile, the nutrient quality of the tissues, and the physiological condition of the insect all play a part (Edwards and Wratten, 1980). Thus an insect may spend a significant time in sampling and rejecting plant tissue before it finally settles to feed. Furthermore, many more insects typically move their site of feeding between feeding bouts, and thus are continuously searching for food of high quality, in terms of both nutrient value and secondary chemistry (Schultz, 1983).

It is less easy to generalize about the characteristics of fungi which influence the threat they pose to plants, though there are some obvious contrasts with insects (Table 6.1). Fungi are not mobile, and do not detect the quality of their food in the same way as insects. Initial establishment is usually by the germination of a spore, and it is largely a matter of chance whether or not the spore lands upon adequate host plant tissues. The opportunities for food 'selection' are thus more restricted, and occur only in the sense that a mycelium, once established, develops most rapidly into those parts of its host which are of the highest quality. Another difference from insects is that fungi are unable to escape from unfavourable conditions of desiccation or high temperatures, and thus may not be able to exploit plant tissues where these conditions regularly occur.

Although many fungal pathogens are characterized by short life cycles and high intrinsic rates of increase, others are a threat to plants precisely because they have the opposite characteristics and are long lived and develop slowly. The mycelial growth form provides the ability to accumulate nutrients in short supply as the hyphae penetrate into fresh substrate and withdraw materials from old hyphae. Indeed, many important fungal pathogens, such as those which attack the wood of living trees, have notably low requirements for mineral nutrients such as nitrogen, and have a remarkable ability to adapt the nutrient content of their mycelium to that of the substrate. For example, in culture experiments the nitrogen content of mycelium of *Polyporus versicolor* varied from 0.1 % when grown on a medium with a C : N ratio of 1600 : 1 to 8.2 % at a C : N ratio of 4 : 1 (Levi and Cowling, 1969). Thus, low nutritional quality of the host plant does not represent the same barrier to fungal attack that it does to insect infestation. Another important contrast between the threat posed by fungi and insects is that the effect of fungal pathogens is commonly systemic because it interferes with systemic processes such as phloem transport or water movement. In contrast,

Table 6.2. *Examples of defensive systems which appear to be particularly effective against either insects or micro-organisms*

Defensive System	Example
Insects	
Deterrents	Aphid alarm pheromones in glandular trichomes, e.g. potato
Physical constraints	Hooked trichomes, e.g. *Phaseolus*, *Passiflora*
	Resins in glandular trichomes producing gummosis, e.g. certain wild potatoes
Toxins	
Constitutive	Active compounds concentrated in vulnerable foliage, e.g. alkaloids in young leaves of *Prosopis*
Induced	Rapid systemic induction of highly toxic compounds, e.g. cucurbitacins in *Cucurbita moschata*
Attractants for predators	Extrafloral nectaries, domatia of ant plants
Fungi	
Barriers to initial penetration	
Structural	Waxy cuticle, thick epidermis and/or hypodermis
Chemical	Tannins/flavanoids in epidermal and outer cortical cells
Barriers induced by infection or wounding	Resins and gums, e.g. most conifers, cork, wound lignin, callose, tyloses, papillae, etc.
Locally induced toxins	Phytoalexins
Systemically induced anti-fungal agents	Pathogenesis related proteins

insect damage does not normally have systemic consequences, and the impact of attack by insects is likely to be the cumulative effects of local defoliation.

6.5.2 *Defensive systems against insects and micro-organisms*

Because of the fundamental differences between insects and fungi, there are likely to be some ecological circumstances in which one group or the other represents the greater selective pressure upon plants. It is often not possible to assign a particular defensive role to a plant substance or structure, but there are certain features of plants which are demonstrably effective against a certain type of attack, and we are probably justified in considering them as specialized defences. Table 6.2 presents some examples.

Defensive systems against insects often reflect and exploit the sensory behaviour and mobility of insects, and are typically located at the plant surface (Juniper and Southwood, 1986). For example, trichomes may provide a physical barrier to movement and may even puncture the cuticle of insects (e.g. those of *Phaseolus vulgaris*; Pillemer and Tingey, 1976). Glandular trichomes may produce secretions which immobilize insects (e.g. the glandular hairs of certain wild potato varieties; Gibson, 1971), or produce secondary substances which act as feeding deterrents or alarm pheromones. All of these systems represent defences which the insect encounters at an early stage of its contact with the plant. In addition, toxic secondary substances are likely to be concentrated in tissues, such as young foliage, which are of high nutritional value to the insect, and of particular importance to the plant.

In some species the production of toxic secondary compounds is induced by wounding (Raupp and Tallamy, 1991). If these compounds are to be an effective defence against the insects feeding on the plant, the induction must be relatively rapid, and occur in tissues remote from the site of damage. Examples of rapidly induced changes which have these characteristics include the synthesis of cucurbitacins in *Cucurbita* spp. (enhanced levels of cucurbitacins are synthesized within 3 h of damage; Tallamy and Krischik, 1991), and proteinase inhibitors in tomato and other members of the Solanaceae (Green and Ryan, 1972; see also Chapter 3).

The first line of defence against fungi is through barriers to penetration, both physical and chemical. Thus, in anti-fungal defensive systems, secondary compounds are likely to be concentrated where attack is most likely, and not necessarily in the tissues of greatest value to the plant. The high concentrations of flavonoids and tannins which commonly occur in the epidermal cells of leaves, and in the outer cortical tissues of stems, appear to represent defences of this kind.

Induced defensive systems against fungal attack are likely to be of a different character from those which are effective against insects. For example, the phytoalexin response, which involves the *de novo* synthesis of highly toxic substances at the point of infection, is unlikely to affect insects significantly because of their mobility. Another characteristically anti-fungal defensive system is the induced production of a structural or chemical barrier between healthy and infected tissue (Ride, 1983); the time-scale of this response is such that it would have no significance for a mobile, foraging organism such as an insect.

Because of their nutritional requirements, insects present their greatest threat to plants in nutrient-rich, productive environments (Edwards, 1989), and their attack is focused upon tissues of high nutritional value (i.e. foliage rather than wood). Plants characteristic of such conditions are usually fast growing and have relatively short-lived, soft leaves with high concentrations of nitrogen and other nutrients (Grime, 1979). Typical examples of such species from the British flora include *Urtica dioica*, *Solanum dulcamara*, *Epilobium hirsutum* and *Humulus lupulus*. Where conditions are difficult for plant growth, for example because of drought, low temperatures or low fertility, plants typically have long-lived leaves which must be resistant to the range of hazards (including micro-organisms) that they encounter over an extended life span (Grubb, 1986). Such foliage may be highly resistant to grazing because it is tough or of low nutrient content, though such resistance does not necessarily represent an evolved defence against herbivores (i.e. it is *neutral* resistance as opposed to *defensive* resistance; Edwards, 1989).

Micro-organisms pose a threat to plants in a much wider range of ecological conditions, and attack is not restricted to particular organs such as leaves. The predominant defensive systems may therefore be expected to vary widely, both between the different organs of a plant, and between species of differing habitats. In general, there appears to be increasing investment in anti-microbial defence in progressively longer-lived tissues. Thus, the nutrient-rich foliage of plants in productive environments is likely to exhibit rapidly induced defences (e.g. phytoalexins, papillae), effective against fast growing, virulent pathogens. In contrast, in leaves of slow growing species, there is a greater emphasis upon constitutive defences, particularly in the form of barriers to infection. The same emphasis is seen in woody tissues, which often have very high concentrations of secondary substances with anti-microbial activity (Hillis, 1962, 1975). Furthermore, wood of long-lived trees is physically stronger and has greater chemical protection against decay-inducing micro-organisms (Grubb, 1992) than that of short-lived species.

6.6 Conclusions

The ideas of Bu'Lock and Gottlieb about the mechanisms by which secondary metabolites arise in plants have important implications for our understanding of their role in interactions between plants and other organisms. Although the evolutionary forces leading to the elaboration of novel biosynthetic pathways may be largely unrelated to biotic factors, the metabolites that are produced represent, in effect, a resource upon which natural selection may act. It is this point which marks a fundamental difference of approach from that of the defence paradigm. We need no longer regard all secondary metabolites as defences for which an adaptive explanation must be sought in terms of protection against some herbivore, pathogen or competitor, whether past or present. Instead we may expect to find that natural selection has taken advantage of particular metabolic pathways to elaborate defensive systems.

The question that we ask about the allocation of resources to defence is thus subtly changed. Instead of seeking a general theory of plant defence which can explain the types and quantities of secondary compounds produced by plants, we need to investigate the types of chemical defensive systems that have evolved, and the circumstances which have favoured such evolution. There can be no doubting that both insects and fungi do represent major selective pressures upon plants, though the relative importance of these pressures varies in different environments. The capacity of insects to exploit plant material is restricted by their nutritional requirements, and they pose their greatest threat to fast-growing plants with nutrient-rich foliage. These plants typically exhibit defensive systems which appear to be clearly targeted against insects. Slow-growing species of more difficult environments have a higher neutral resistance, but do not necessarily have highly evolved anti-insect defences. In contrast, micro-organisms represent a more pervasive threat, and anti-microbial defence is probably universal, though investment in such defences is greatest in long-lived tissues.

References

Berenbaum, M. (1983) Coumarins and caterpillars: a case for coevolution. *Evolution* **37**, 163–179.

Bernays, E.A. (1981) Plant tannins and insect herbivores: an appraisal. *Ecol. Entomol.* **6**, 353–360.

Bernays, E.A., Cooper Driver, G. and Bilgener, M. (1989) Herbivores and plant tannins. *Adv. Ecol. Res.* **19**, 263–302.

Bu'Lock, J.D. (1980) Mycotoxins as secondary metabolites. In: *The Biosynthesis of Mycotoxins* (ed. P.S. Steyn). Academic Press, New York, pp. 1–16.

Coley, P.D. (1986) Costs and benefits of defense by tannins in a neotropical tree. *Oecologia* **70**, 238–241.

Coley, P.D., Bryant, J.P. and Chapin, F.S. (1985). Resource availability and plant antiherbivore defense. *Science* **230**, 895–899.

Edwards, P.J. (1989) Insect herbivory and plant defence theory. In: *Toward a More Exact Ecology* (eds P.J. Grubb and J.B. Whittaker). Blackwell Scientific Publications, Oxford, pp. 275–297.

Edwards, P.J. and Wratten, S.D. (1980) *Ecology of Insect–Plant Interactions.* Edward Arnold, London.

Ehrlich, P.R. and Raven, P.H. (1964) Butterflies and plants: a study in plant coevolution. *Evolution* **18**, 586–608.

Fagerstrom, T. (1989) Anti-herbivore chemical defense in plants: a note on the concept of cost. *Am. Nat.* **133**, 281–287.

Farrell, B.D., Dussourd, D.E. and Mitter, C. (1991) Escalation of plant defense: do latex and resin canals spur plant diversification? *Am. Nat.* **138**, 881–900.

Feeny, P.P. (1970) Seasonal changes in oak-leaf tannins and nutrients as a cause of spring feeding by winter moth caterpillars. *Ecology* **51**, 565–581.

Feeny, P. (1976) Plant apparency and chemical defenses. *Recent Adv. Phytochem.* **10**, 1–41.

Fox, L.R. (1981) Defense and dynamics in plant-herbivore systems. *Am. Zool.* **21**, 853–864.

Fraenkel, G.S. (1959) The raison d'être of secondary plant substances. Science **129**, 1466–1470.

Gibson, R.W. (1971) Glandular hairs providing resistance to aphids in certain wild potato species. *Ann. Appl. Biol.* **68**, 113–119.

Gottlieb, O.R., (1989) The role of oxygen in phytochemical evolution towards diversity. *Phytochem.* **28**, 2545–2558.

Gottlieb, O.R. (1990) Phytochemicals: differentiation and function. *Phytochem.* **29**, 1715–1724.

Green, T.R. and Ryan, C.A. (1972) Wound-induced proteinase inhibitor in plant leaves: a possible defense mechanism against insects. *Science* **175**, 776–777.

Grime, J.P. (1979) *Plant Strategies and Vegetation Processes.* Wiley, Chichester.

Grubb, P.J. (1986) Sclerophylls, pachyphylls and pycnophylls: nature and significance of hard leaf surfaces. In: *Insects and the Plant Surface* (eds B. Juniper and T.R.E. Southwood). Edward Arnold, London, pp. 137–150.

Grubb, P.J. (1992) A positive distrust in simplicity: lessons from plant defences and from competition among plants and among animals. *J. Ecol.* **80** (in press).

Gulman, S.L. and Mooney, H.A. (1986) Costs of defense and their effect on plant productivity. In: *On the Economy of Plant Form and Function* (ed. T.J. Givnish). Cambridge University Press, Cambridge, pp. 681–698.

Harborne, J.B. (1988) *Introduction to Ecological Biochemistry*, 3rd edition. Academic Press, London.

Haslam, E. (1988) Plant polyphenols (syn. vegetable tannins) and chemical defense – a reappraisal. *J. Chem. Ecol.* **14**, 1789–1806.

Heltmann, E. (1975) Functions of steroids in plants. *Phytochem.* **14**, 891–902.

Hillis, W.E. (1962) *Wood Extractives.* Academic Press, New York.

Hillis, W.E. (1975) Secondary changes in wood. In: *The Structure, Biosynthesis and Degradation of Wood* (eds F.A. Loewus and V.C. Runeckles). Plenum Press, New York, pp. 247–309.

Janzen, D.H. (1969) Coevolution. *Science* **165**, 415.

Jermy, T. (1984) Evolution of insect/host plant relationships. *Am. Nat.* **124**, 609–630.

Juniper, B. and Southwood, T.R.E. (1986) *Insects and the Plant Surface*. Edward Arnold, London.

Kevan, P.G., Chaloner, W.G. and Savile, D.B.O. (1976) Interrelationships of early terrestrial arthropods and plants. *Palaeontology* **18**, 391–417.

Levi, M.P. and Cowling, E.B. (1969) Role of nitrogen in wood deterioration. VII Physiological adaptation of wood-destroying and other fungi to substrates deficient in nitrogen. *Phytochem.* **59**, 460–468.

Mansfield, J.W. (1983) Antimicrobial compounds. In: *Biochemical Plant Pathology* (ed. J.A. Callow). Wiley, Chichester, pp. 237–266.

Marquis, R.J. and Alexander, H.M. (1992) Evolution of resistance and virulence in plant-herbivore and plant-pathogen interactions. *Tree* **7**, 126–129.

Mattson, W.J. (1980) Herbivory in relation to plant nitrogen content. *Ann. Rev. Ecol. Syst.* **11**, 119–161.

Mitter, C., Farrell, B. and Futuyma, D.J. (1991) Phylogenetic studies of insect-plant interactions: insights into the genesis of diversity. *Tree* **6**, 290–293.

Muller, C.H. (1969) The "Co-" in coevolution. *Science* **164**, 197.

Niemeyer, H.M. and Perez, F.J. (1987) Hydroxamic acids from Gramineae: their role in aphid resistance and their mode of action. In: *Insects–Plants* (eds V. Labeyrie, G. Fabres and D. Lachaise). W. Junk, Dordrecht, pp. 49–52.

Pillemer, E.A. and Tingey, W.M. (1976) Hooked trichomes: a physical plant barrier to a major agricultural pest. *Science* **193**, 482–484.

Pirozynski, K.A. (1976) Fossil fungi. *Ann. Rev. Phytopathol.* **14**, 237–246.

Raupp, M.J. and Tallamy, D.W. (1991) *Phytochemical Induction by Herbivores*. Wiley, New York.

Ride, J.P. (1983) Cell walls and other structural barriers in defence. In: *Biochemical Plant Pathology* (ed. J.A. Callow). Wiley, Chichester, pp. 215–236.

Scalbert, A. (1991) Antimicrobial properties of tannins. *Phytochem.* **30**, 3875–3883.

Schultz, J. (1983) Impact of variable plant defensive chemistry on susceptibility of insects to natural enemies. In: *Plant Resistance to Insects* (ed. A. Hedin). American Chemical Society Symposium No. **208**, pp. 37–55.

Shutt, D.A. (1976) The effects of plant oestrogens on animal reproduction. *Endeavour* **35**, 110–113.

Sporne, K.R. (1980) A re-investigation of character correlations among dicotyledons. *New Phytol.* **85**, 419–449.

Strong, D.R., Lawton, J.H. and Southwood, T.R.E. (1984) *Insects on Plants: Community Patterns and Mechanisms*. Blackwell Scientific Publications, Oxford.

Swain, T. (1975) Evolution of flavonoid compounds. In: *The Flavonoids* (eds J.B. Harborne, T.J. Mabry and H. Mabry). Chapman and Hall, London, pp. 1096–1129.

Tallamy, D.W. and Krischik, V.A. (1989) Variation and function of cucurbitacins in *Cucurbita*: an examination of current hypotheses. *Am. Nat.* **133**, 766–786.

Tuomi, J., Niemela, P., Chapin, F.S., Bryant, J.P. and Siren, S. (1988) Defensive responses of trees in relation to their carbon/nutrient balance. In: *Mechanisms of Woody Plant Defenses against Insects: Search for Pattern* (eds W.J. Mattson, J. Levieux and C. Bernard-Dagan). Springer, New York, pp. 57–72.

Tuomi, J., Niemela, P., Haukioja, E., Siren, S. and Neuvonen, S. (1984) Nutrient stress: an explanation for plant anti-herbivore responses to defoliation. *Oecologia* **61**, 208–210.

Zucker, W.V. (1983) Tannins: does structure determine function? An ecological perspective. *Am. Nat.* **121**, 335–365.

7

Photosynthesis: cellular and tissue aspects in diseased leaves

J.D. Scholes

7.1 Introduction

Biotrophic pathogens of leaves (the rusts and powdery and downy mildews) may cause a transient stimulation of the rate of photosynthesis but, in most host/pathogen interactions, both net and gross photosynthesis then decline and chlorophyll is lost from the tissue as infection progresses (Magyarosy *et al.*, 1976; Owera *et al.*, 1981; Ahmad *et al.*, 1983; Scholes and Farrar, 1986; Farrar and Lewis, 1987; Scholes *et al.*, 1992). In addition, biotrophs generally increase the rate of respiration and alter the amounts and translocation of carbohydrates in infected leaves (for reviews see Whipps and Lewis, 1981, and Farrar, this volume).

Many mechanisms, summarized in Table 7.1, have been proposed to explain the reduction in the rate of photosynthesis. Briefly, they can be divided into physical factors, such as alterations in the pathway of diffusion of CO_2 into and within a leaf, and cellular or metabolic changes. The latter include an increase in host and/or fungal respiration (this would reduce the rate of net photosynthesis), direct fungal inhibition of photosynthetic electron transfer by the production of a toxin or by the selective destruction of components of the electron transport chain, a reduction in the amount and/or activity of ribulose-1,5-bisphosphate carboxylase/oxygenase (Rubisco), destruction of photosynthetic machinery, feedback inhibition of photosynthesis following carbohydrate accumulation and fungal sequestration of inorganic phosphate (P_i) from the host cytoplasm. The last two mechanisms would lead to a reduction in the recycling of P_i to the chloroplast with the consequence that ATP synthesis and 3-phosphoglycerate (3-PGA) reduction would become restricted, starch synthesis stimulated and the rate of photosynthesis reduced. However, it is not possible from most of these studies to identify the primary and secondary causes of the inhibition of the rate of photosynthesis. Some studies suggest that pathogens act specifically to

Table 7.1. *Mechanisms proposed to account for reduction in rate of photosynthesis in leaves infected by fungal biotrophic pathogens*

Mechanism	Host/pathogen system	Reference
(A) Physical mechanisms		
Alterations in the diffusion of CO_2 into and within the leaf	Powdery mildew of sugar beet	Gordon and Duniway, 1982a
	Powdery mildew of wheat	Ayres, 1976
	Powdery mildew of soybean	Mignucci and Boyer, 1979
	Powdery mildew of oak	Hewitt and Ayres, 1975
	(Review article)	Ayres, 1981
(B) Cellular/metabolic mechanisms		
1. Increased host and/or fungal respiration	Brown rust of barley	Scott and Smillie, 1966; Daly, 1976; Walters and Ayres, 1984; Owera *et al.*, 1981; Scholes and Farrar, 1986
	Stem rust of wheat	Samborski and Shaw, 1956; Mitchell, 1979
	Rust of bean	Raggi, 1980
	Powdery mildew of wheat	Zulu *et al.*, 1991a; Wright, 1992
	Powdery mildew of barley	Farrar and Rayns, 1987
	Powdery mildew of pea	Ayres, 1976
2. Altered solute relations within host cells	Rust of bluebell	Scholes and Farrar, 1985
	Stem rust of wheat	Whitney *et al.*, 1962
3. Inhibition of photosynthetic electron transfer		
(i) Non-cyclic photophosphorylation	Powdery mildew of sugar beet	Magyarosy *et al.*, 1976
	Rust of broad bean	Montalbini and Buchanan, 1974
	Rust of bluebell	Scholes and Farrar, 1985
(ii) Loss of components of the photosynthetic electron transfer chain	Powdery mildew of sugar beet	Magyarosy and Malkin, 1978

4. Reduction in the amount and/or activity of enzymes of the Calvin cycle		
(i) Rubisco	Powdery mildew of barley	Walters and Ayres, 1984; Higgins *et al.*, 1985; Scholes *et al.*, 1992
	Powdery mildew of wheat	Wright, 1992
	Powdery mildew of sugar beet	Gordon and Duniway, 1982b
	Rust of leek	Roberts and Walters, 1988
	Late blight of potato	Kombrink and Hahlbrock, 1990
	Colletotrichum lagenarium on melon	Ranty *et al.*, 1987; Roby *et al.*, 1988
(ii) NADP-GAPDH	Powdery mildew of barley	Walters and Ayres, 1984; Scholes *et al.*, 1992
	Powdery mildew of wheat	Wright, 1992
(iii) Stromal fructose-1,6-bisphosphatase	Powdery mildew of barley	Scholes *et al.*, 1992
	Powdery mildew of wheat	Wright, 1992
(iv) 3-Phosphoglycerate kinase	Powdery mildew of barley	Walters and Ayres, 1984
5. Accumulation of carbohydrate causing		
(i) end-product inhibition	Powdery mildew of barley	Scholes *et al.*, 1992
	Powdery mildew of wheat	Wright, 1992
(ii) down-regulation of gene expression	Powdery mildew of barley	Scholes *et al.*, 1992
	Powdery mildew of wheat	Wright, 1992
6. Fungal sequestration of inorganic phosphate from the host cytoplasm	(Hypothesis only)	Whipps and Lewis, 1981

disrupt photosynthetic electron transfer whilst others (in many cases using the same host/pathogen system) indicate that the primary effect of the pathogen is to decrease the amount of photosynthetic proteins (Table 7.1).

There are many reasons for these apparent contradictions. Three, in particular, can be singled out. Firstly, most studies have examined the effect of a particular pathogen on only one component of the photosynthetic process and, therefore, it is impossible to sequence events and to identify primary mechanisms. Secondly, it cannot be assumed that all biotrophic pathogens will act mechanistically in a similar manner. Powdery mildews are ectoparasites with haustoria confined to the epidermal cells. Thus, carbohydrates and other nutrients must pass through these cells prior to uptake by the fungus. In contrast, rust and downy mildew pathogens form haustoria in epidermal, mesophyll and bundle sheath cells and therefore have a more direct access to pools of carbohydrates. In addition, an extensive intercellular mycelium allows uptake of nutrients from the apoplast. Such differences in growth form within the leaf may have profound consequences for the mechanism of inhibition of photosynthesis. Thirdly, diseased leaves are extremely heterogeneous; they consist both of cells directly invaded by the fungus and of cells that are not invaded but which are modified by its presence. This is particularly true of rust diseases. In bluebell leaves infected with *Uromyces muscari* (Scholes and Farrar, 1985) and leek leaves infected with *Puccinia allii* (Roberts and Walters, 1988), both net and gross photosynthesis were reduced within pustules, but in uninvaded areas of the infected leaf they were similar to control leaves. In contrast, in barley leaves infected with brown rust (*Puccinia hordei*), the reverse was true. Although brown rust caused an overall reduction in the rate of photosynthesis, autoradiographic studies showed that the decline was attributable to a severe reduction in the rate of photosynthesis in regions between pustules and that, in fungally invaded areas, it was higher than in control leaves (Scholes and Farrar, 1986). The rate of photosynthesis measured from the whole leaf will, therefore, depend in part upon the density of infection.

The rate of photosynthesis is controlled by different factors depending upon the experimental conditions in which it is measured (Stitt and Quick, 1989; Stitt *et al.,* 1991). For example, in limiting irradiance, it is determined by the efficiency of light harvesting and use. In saturating light and ambient CO_2, photosynthesis is determined, to a large extent, by Rubisco (Stitt *et al.,* 1991), whereas, in saturating light and saturating CO_2, it is mainly determined by the rate of end-product synthesis (Leegood and Furbank, 1986; Stitt and Quick, 1989). Thus, careful selection of experimental conditions will allow major effects on the partial processes of photosynthesis to be identified. However, the experimental conditions chosen to measure the effect of pathogens on the rate of photosynthesis vary between studies. In many cases, such differences will influence the results in ways which have rarely been taken into account when comparing studies. Finally, the conclusions drawn from many studies now need to be re-examined in the light of new ideas concerning the regulation of photosynthesis.

Recently, two studies tried to elucidate the precise sequence of events and mechanisms underlying the reduction in the rate of photosynthesis in cereal leaves following infection with powdery mildew, *Erysiphe graminis* (Scholes *et al.*, 1992; Wright, 1992).

Simultaneous measurements of the rate of photosynthesis and of chlorophyll-*a* fluorescence (to provide information on the functioning and regulation of photosystem 2 (PSII)), and freeze clamping of leaves (to allow analysis of enzymes and intermediates of the Calvin cycle, sucrose and starch synthesis), throughout the infection cycle, suggested that the inhibition of the rate of photosynthesis in infected leaves occurred primarily as a result of a loss of activity and/or amount of photosynthetic enzymes of the Calvin cycle. This followed an increase in the activity of acid invertase and, concomitantly, an accumulation of glucose, fructose and sucrose. These authors suggested that the down-regulation of the Calvin cycle occurred either as a result of end-product inhibition or more probably as a direct effect of carbohydrates on the expression of genes encoding photosynthetic enzymes (Scholes *et al.,* 1992; Wright, 1992). Electron transfer was then down-regulated in response to a decreased demand for ATP and NADPH, and photoinhibition and accelerated loss of chlorophyll from the leaf occurred .

This hypothesis (Figure 7.1) is now considered in detail in order to provide a framework for discussion of other results often with other host–pathogen interactions. Firstly, current ideas concerning the mechanisms of feedback or 'sink' regulation of photosynthesis are reviewed briefly.

7.2 Regulation of photosynthesis

The photosynthetic system can be considered as a branched metabolic pathway with one input branch mediating the capture of CO_2 from the atmosphere and the other the input of quantum energy for the reduction of CO_2 to sucrose, the major product of photosynthesis (Woodrow and Berry, 1988). For photosynthesis to operate at maximum efficiency under a given set of conditions, the two input fluxes must conform to a specific stoichiometry so that they do not exceed the capacity of downstream metabolic pathways (e.g. for sucrose and starch synthesis) to sustain the fluxes. Since both photosynthetic electron transfer and the Calvin cycle contain steps which are essentially irreversible, there must be regulatory mechanisms to co-ordinate the rates of these reactions. When light energy is absorbed by chlorophyll molecules, the quantum energy is either used for photochemistry within a few picoseconds or dissipated as heat or fluorescence (for reviews see Horton and Bowyer, 1990; Krause and Weis, 1991). Once charge separation has occurred, electron transfer is largely irreversible. The input of CO_2 is mediated by a series of reversible diffusion processes (Farquhar and Sharkey, 1982) and the largely irreversible reactions catalysed by Rubisco. Thus, for steady-state photosynthesis to be maintained, three sets of regulatory sequences are required.

Firstly, at high irradiance when the input of light energy exceeds the capacity of Rubisco to fix CO_2, a regulatory sequence must ensure that excess energy is dissipated non-photochemically as heat or fluorescence (Foyer *et al.*, 1990; Horton and Bowyer, 1990; Krause and Weis 1991). Secondly, at low irradiances, the activity of Rubisco must be modulated so that it matches the input of light energy. This modulation mostly involves changes in the activation of Rubisco (by Rubisco activase) or in the proportion of carbamylated to decarbamylated binding sites (for a review see

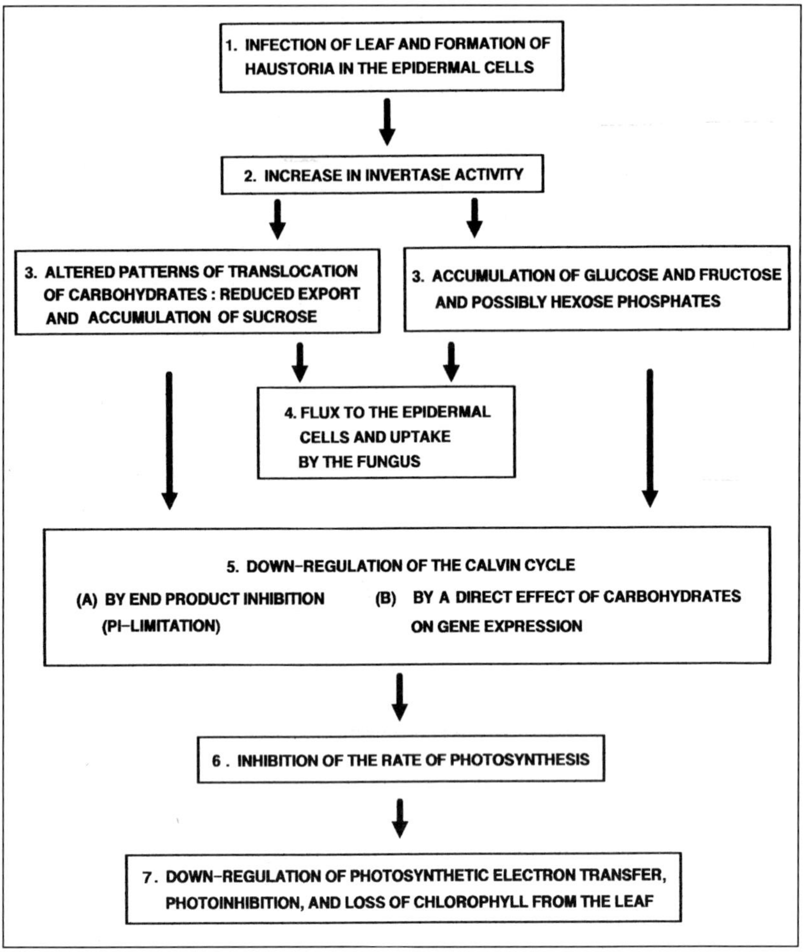

***Figure* 7.1**. *Hypothetical sequence of changes in the carbohydrate and photosynthetic metabolism of cereal leaves infected with powdery mildew.*

MacDonald and Buchanan, 1990). Finally, if the maximum activity of reactions involved in product synthesis is reached, then a sequence that can reduce both the input of light energy and Rubisco activity must be activated. Aspects of this last sequence are considered in more detail.

7.2.1 *Feedback 'sink' regulation of photosynthesis*

Photosynthesis is reduced when export from the (source) leaf is inhibited. This may happen if the leaf is detached (Geiger, 1976), or if, the importing 'sink' leaf or organ is removed or has its metabolism impaired, for example by herbivory, or if, as found recently, plants are transformed such that they over-express yeast invertase in their cell

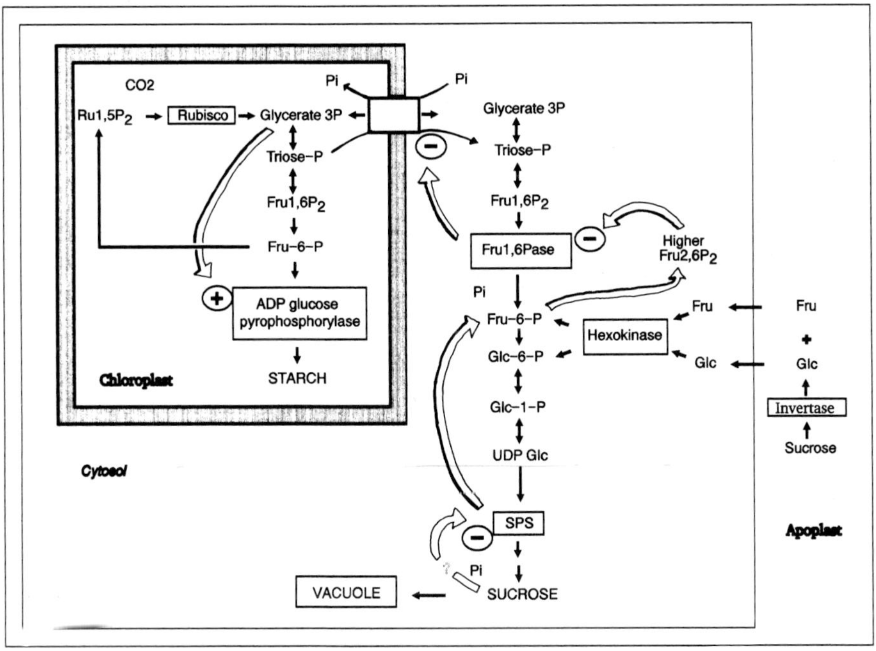

Figure 7.2. *Feedback regulation of photosynthetic sucrose synthesis. Abbreviations in the text. Arrows represent the regulatory sequence of events. Diagram modified from Stitt (1990a).*

walls (von Schaewen *et al.,* 1990; Stitt *et al.*, 1991). This 'sink' regulation of photosynthesis is frequently accompanied by the accumulation of carbohydrates in the leaf (Bagnall *et al.*, 1989; Sawada *et al.*, 1989). Accumulation could lead to an inhibition of photosynthesis by various mechanisms including, (a) a direct effect via P_i-limitation, because feedback inhibition restricts the rate of sucrose or starch synthesis (Herold, 1980; Walker and Sivak, 1986; Foyer, 1987), or (b) a more indirect and long-term effect via a decrease in the amount of proteins required for photosynthesis (Sheen, 1990).

7.2.2 *P_i-limitation of photosynthesis*

During photosynthesis, triose phosphates (triose-P) are exported to the cytosol and used in the synthesis of sucrose. P_i is released and returned to the chloroplast via a strict counter-exchange mediated by the phosphate translocator (Stitt *et al.*, 1987). The rate of sucrose synthesis is regulated by feedforward mechanisms which co-ordinate it with the rate of photosynthesis and by feedback mechanisms which allow the rate of sucrose synthesis to be decreased (Figure 7.2). Several enzymes are involved in these regulatory loops including the cytosolic fructose-1,6-bisphosphatase (Fru1,6Pase) and sucrose phosphate synthase (SPS). Fru1,6Pase is regulated (inhibited) by the signal metabolite fructose-2,6-bisphosphate (Fru2,6P_2) (Stitt, 1990a,b).

As the rate of photosynthesis increases, more triose-P will be exported from the chloroplast in exchange for P_i, with the result that the cytosolic PGA : P_i ratio increases. This causes a decline in the concentration of Fru2,6P_2 and a stimulation of cytosolic Fru1,6Pase (Stitt, 1990a). This enzyme is also stimulated by an increasing concentration of its substrate, Fru1,6P_2 (Herzog *et al.*, 1984). Increased flux through this pathway will stimulate SPS via an increase in the glucose 6-phosphate (Glc6P): P_i ratio (Doehlert and Huber, 1983, 1984). This feedforward regulation is counteracted by a feedback loop as follows.

When the rate of sucrose synthesis exceeds export and sucrose accumulates in leaves, triose-P is not exported but is retained for starch synthesis in the chloroplast. The mechanisms involved in the feedback regulation of sucrose synthesis are not yet fully understood. It has been suggested that accumulating sucrose directly inhibits enzymes involved in its synthesis but evidence is not convincing (Gerhardt *et al.*, 1987). More recently Huber *et al.* (1989) and Huber and Huber (1991) have proposed that sucrose may indirectly trigger protein modification or turnover. In whatever way it occurs, a reduction in the activity of SPS will lead to an accumulation of fructose 6-phosphate (Fru6P) in the cytosol (Figure 7.2). This in turn will lead to an inhibition of fructose-2,6-bisphosphatase (Fru2,6P_2ase), resulting in an increase of Fru2,6P_2. The increased Fru2,6P_2 pool will then inhibit Fru1,6Pase and cause an accumulation of triose-P. The decreased recycling of P_i to the chloroplast will restrict ATP synthesis and, in turn, restrict the reduction of PGA to triose-P within the chloroplast because phosphoglycerate kinase is very sensitive to decreasing concentrations of ATP. An increase in the PGA : P_i ratio within the chloroplast will strongly activate ADP-glucose pyrophosphorylase and so stimulate starch synthesis (Preiss, 1982). If P_i limitation is severe, it will eventually lead to an inhibition of the rate of photosynthesis, as has been demonstrated by growing barley, spinach and soybean plants at low P_i (Dietz and Foyer, 1986; Foyer and Spencer, 1986) and by feeding P_i-sequestering agents to leaves (Herold and Lewis, 1977; Leegood *et al.*, 1988).

Foyer (1987) suggested that an increasing cytosolic sucrose concentration may be accompanied by a 'leakage' or transport of sucrose from the symplast to the apoplast, and that this would facilitate the down-regulation of the system. The apoplast is an unfavourable environment for sucrose since the cell walls contain substantial invertase activity. Sucrose would therefore be hydrolysed to glucose and fructose which would re-enter the cytosol and become phosphorylated by cytosolic hexokinase. Two immediate effects would be an increased demand for ATP and an accumulation of Fru6P which would then trigger the down-regulation of photosynthesis (Figure 7.2).

7.2.3 *Direct effect of accumulation of carbohydrates on the expression of photosynthetic genes*

Carbohydrates may play another role in the feedback regulation of photosynthesis, namely the switching off of photosynthetic genes. Using isolated maize mesophyll protoplasts and a transient expression assay, Sheen (1990) showed that the transcriptional activity of seven photosynthetic gene promotors was repressed specifically and in a co-ordinated manner by high concentrations of sucrose and glucose. Analysis of deleted, mutated and hybrid promoters indicated that the sugars inhibited the activity of distinct positive upstream regulatory elements.

Stitt *et al.* (1991) and von Schaewen *et al.* (1990) reported that apoplastic phloem loading and export were inhibited when yeast-derived invertase was expressed in the cell walls of tobacco and *Arabidopsis* plants. The accumulation of starch, sucrose, glucose and fructose was accompanied by a progressive inhibition of photosynthesis and a stimulation of respiration. Analysis of the kinetics of chlorophyll fluorescence and measurements of both metabolites and enzyme activities did not provide support for the hypothesis that photosynthesis was inhibited as a result of end-product accumulation and P_i-limitation. Rather, it was inhibited by a decrease of several photosynthetic enzymes including Rubisco. Secondly, analogous results were obtained when mature spinach leaves were supplied with glucose via the transpiration stream (Krapp *et al.,* 1991). Carbohydrates accumulated, also, when plants were exposed to enhanced CO_2 or to source/sink manipulations. The accumulation of carbohydrate was often accompanied by a decrease of Rubisco (Wong, 1979; Yelle *et al.*, 1989; Besford *et al.,* 1990) and other enzymes of the Calvin cycle (Besford, 1990). In all of these studies, the precise mechanism that 'triggered' the loss of photosynthetic proteins is unknown.

7.3 Inhibition of photosynthesis in cereal leaves infected with powdery mildew: a hypothesis

7.3.1 *Increase in acid invertase and accumulation of carbohydrates*

As haustoria of powdery mildews developed in epidermal cells of compatible barley and wheat leaves, there was a rapid increase in the activity of acid invertase 3–7 days after inoculation (dai) (Figure 7.3a). Enhanced invertase activity has previously been reported for many rust, powdery and downy mildew infections (Long *et al.,* 1975; Billet *et al.,* 1977; Callow *et al.,* 1980; Greenland and Lewis, 1981, 1983; Krishnan and Pueppke, 1988). The precise origin and location of the increase in invertase activity remains controversial as both the host and pathogen possess soluble and insoluble invertases. Therefore, it is difficult to establish whether the stimulation in invertase activity is due to an activation or an increase in the amount of host protein, to a fungal invertase (or α-glucosidase) (Donaldson and Jorgenson, 1988) situated in the extra-haustorial matrix, or to a secretion of a fungal enzyme into the host tissue.

In the studies of cereal mildews, increased activity of invertase was at least partly of host origin; it was not related linearly to the number of haustoria in the epidermal cells and invertase activity remained high in mesophyll cells of barley leaves following removal of the epidermis and fungal mycelium (Figure 7.3b). Such a view is supported by two recent studies. Sturm and Chrispeels (1990) used a cDNA probe for carrot apoplastic invertase to investigate the expression of this enzyme in carrot roots and leaves infected with the bacterial pathogen *Erwinia carotovora* . Within 1 h of infection, there was a rapid increase in the amount of host mRNA. Benhamou *et al.* (1991) used an antiserum raised against a purified deglycosylated carrot cell wall invertase to study, by immunogold labelling, the spatial and temporal distribution of the enzyme in

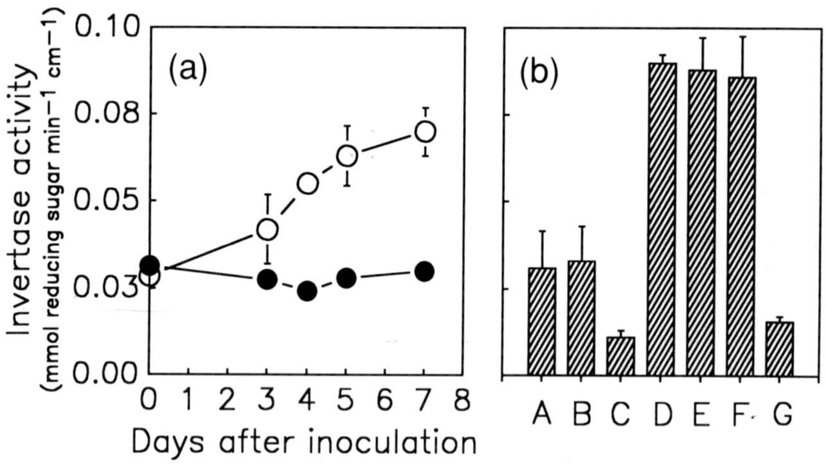

Figure 7.3. *(a) Effect of powdery mildew infection on the activity of invertase in barley leaves; control (●) and infected (○) leaves. (b) Cellular location of the increase in invertase activity in barley leaves infected with powdery mildew (on the abaxial surface only) 4 days after inoculation. (A) Uninfected intact leaf; (B) mesophyll cells of an uninfected leaf, from which the abaxial epidermis had been removed; (C) abaxial epidermis of an uninfected leaf; (D) an infected, intact leaf; (E) an infected leaf assayed after the surface mycelium had been brushed from the leaf; (F) mesophyll cells of an infected leaf, from which the abaxial epidermis (+ mycelium and haustoria) had been removed; (G) the abaxial surface of an infected leaf. Results are the mean ± s.e. of four measurements. From Scholes* et al. *(1992).*

tomato root tissues infected with the necrotrophic fungus, *Fusarium oxysporum*. Within 72 h of inoculation the enzyme accumulated in the cell walls of colonized cells. At present the potential signal from the fungus to the host is unknown although auxin is a potential candidate, in view of thc established effects of auxin on the regulation of invertase activity (Weil and Rausch, 1990) and the fact that biotrophic pathogens are known to produce auxin, or stimulate their host to produce auxin (Pegg, 1981).

An increase in the activity of acid invertase in the mildewed leaves will have several consequences (Figure 7.1). Firstly, it will increase the hydrolysis of sucrose to glucose and fructose leading to an accumulation of these carbohydrates. This occurs in both mildewed wheat and barley leaves (Figure 7.4). It is generally assumed that biotrophic fungi preferentially take up glucose and fructose into the haustorial complexes (Whipps and Lewis, 1981), although uptake of sucrose has also been reported (Manners and Gay, 1983). Secondly, if part of the increase in activity is apoplastic, the loading of sucrose into the phloem will be prevented because, although the movement of sucrose between mesophyll cells is generally accepted to be symplastic, much evidence favours the presence of an apoplastic transport step just prior to the entry of sucrose into the minor phloem veins (Giaquinta, 1983). Decreased export of sucrose from infected leaves has been reported for many biotrophic infections (for reviews see Whipps and Lewis, 1981; Farrar and Lewis, 1987).

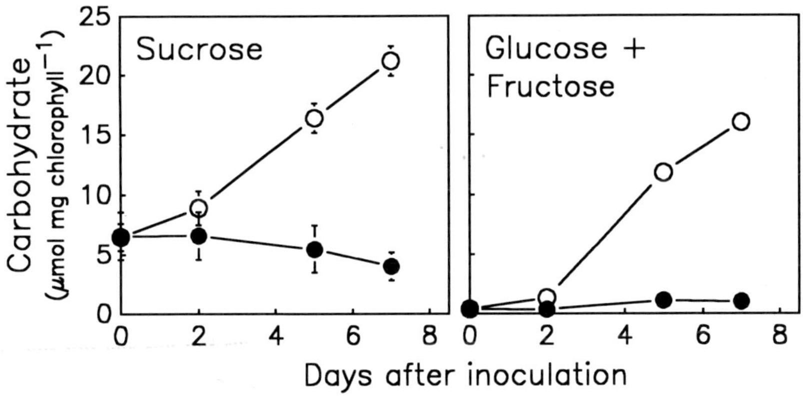

***Figure 7.4**. Effect of powdery mildew on amount of carbohydrate in barley leaves throughout infection. Control (●) and infected (○) leaves. Results are the mean ± s.e. of four measurements. Modified from Scholes* et al. *(1992).*

In both powdery mildew-infected wheat and barley leaves, sucrose accumulated as infection progressed (Figure 7.4). Such an accumulation has also been seen by other research groups working on powdery mildews of cereals (Klecan and Buchanan, 1988; Zulu *et al.*, 1991a) and of vine (Brem *et al.*, 1986). The accumulation of sucrose in the cereal leaves may have resulted from decreased export or increased import of sucrose into infected leaves, or sucrose synthesis may have been maintained at a rate greater than or equal to its hydrolysis. Certainly until 5 dai, sucrose synthesis was maintained to some extent in infected barley and wheat leaves as there was an increase in concentration following a dark to light transition (Scholes *et al.*, 1992) and substantial incorporation of $^{14}CO_2$ into sucrose (Wright, 1992). Note that in most studies of the effect of rust and downy mildew diseases on sucrose, concentration of this carbohydrate is dramatically reduced in the infected leaf when compared to control values (Brem *et al.*, 1986; Gwary, 1988; Tetlow, 1990). The reasons for this apparent difference in sucrose concentration in mildewed and rusted leaves is unknown, although it must be remembered that infected leaves are heterogeneous and detailed studies on the precise location of the changes in sucrose concentration are lacking.

The consequences of the accumulation of sucrose, glucose and fructose on sucrose synthesis and, more generally, on the rate of photosynthesis will depend largely on their distribution between compartments within cells and on fluxes between compartments and between cells. If these carbohydrates are predominantly in the apoplast, vacuole, or in the epidermal cells, photosynthesis will not be affected in the short term. If, however, they accumulate to a high concentration in the cytosol, photosynthesis may be inhibited as a result of end-product inhibition (via P_i-limitation) or, over a longer time period, by a decrease in the synthesis of photosynthetic enzymes.

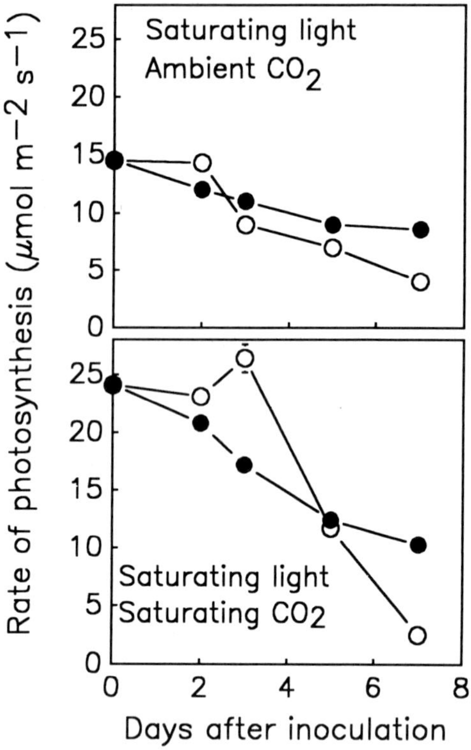

***Figure 7.5**. Effect of powdery mildew on the rate of photosynthesis in barley leaves at saturating light (1000 μmol m^{-2} s^{-1}) and either ambient or saturating CO_2 throughout infection. Control (●) and infected (○) leaves. Results are the mean ± s.e. of four measurements. Modified from Scholes* et al. *(1992).*

7.3.2 *What is the evidence for P_i-limitation of photosynthesis?*

Whipps and Lewis (1981) suggested that the introduction of a fungus into a source leaf might be expected to mimic the effects of P_i-deficiency since biotrophic pathogens act as a sink for P_i, taking up and synthesizing large amounts of polyphosphate within the mycelium. This could reduce the concentration of host cytosolic and thus chloroplastic P_i, causing a reduction in the rate of photosynthesis via the mechanisms described previously, and an increase in the deposition of starch. A major problem in testing this hypothesis and the hypothesis that carbohydrate accumulation limits the availability of P_i, is the difficulty of measuring cytosolic P_i. However, leaves which are P_i-limited generally exhibit a number of physiological characteristics.

Firstly, as mentioned earlier, inhibition of photosynthesis as a result of end-product inhibition is most evident when photosynthesis is measured in conditions of saturating CO_2 and light. In mildewed barley leaves, the rate of photosynthesis progressively decreased as infection developed but the extent of the inhibition was actually less, and occurred later, at high CO_2 and light than at ambient CO_2 (Figure 7.5). In addition, photosynthetic oscillations which can occur under these conditions and which are attributed to P_i-deficiency (Walker and Sivak, 1986) were absent or smaller in amplitude in the infected leaves (Scholes *et al.,* 1990). Secondly, P_i-limitation in leaves causes an accumulation of 3-PGA and ribulose 1,5-bisphosphate ($Ru1,5P_2$), a decrease in the ATP:ADP ratio, and an inactivation of Rubisco (Leegood and Furbank, 1986; Stitt and Quick, 1989). In both mildewed wheat and barley leaves, the amount of PGA decreased (181.7 $\pm$ 14.1 and 239.0 $\pm$ 20.9 nmol mg^{-1} chlorophyll in mildewed and control barley leaves, respectively), while Ru1,5P_2 increased slightly as infection progressed (122.7 $\pm$ 18.2 nmol mg^{-1} chlorophyll in mildewed barley leaves and 98.0 $\pm$ 8.6 nmol mg^{-1} chlorophyll in control tissue). In addition, although the initial activity of Rubisco was less (44.8 $\pm$ 6.9 *vs* 81.9 $\pm$ 14.7 μmol mg^{-1} chlorophyll h^{-1}) in infected than healthy leaves, their activation states (initial/maximal activity) were similar, 34 and 31%, respectively.

A third symptom of P_i-deficiency is an increase in the partitioning of carbohydrate to starch and its subsequent accumulation. In cereal leaves infected with mildew, there was a small increase in starch content (0.8 $\pm$ 0.12 and 0.52 $\pm$ 0.1 μmol mg^{-1} chlorophyll in mildewed and control leaves, respectively). This increase is of similar magnitude to that reported for other studies of mildew infected leaves (Zulu *et al.*, 1991a, b) but it is not as great as that reported for leaves infected with rust pathogens (MacDonald and Strobel, 1970; Ahmad *et al.*, 1984). Also, since the measurements do not distinguish between fungal glucan and host starch, it is unwise to speculate on the origin of the increase. Five days after inoculation of wheat leaves with powdery mildew, incorporation of $^{14}CO_2$ (measured in an O_2 electrode) was reduced in line with the reduction in the rate of photosynthesis. However, the proportion of the fixed CO_2 partitioned into starch was only slightly greater in mildewed than in control leaves (Wright, 1992). A similar result was obtained by Zulu *et al.* (1991b), again in powdery mildew infected wheat leaves.

These data suggest that, although photosynthesis was reduced in infected leaves, this was not caused primarily by P_i-limitation. Such a conclusion is supported by two other lines of evidence.

(i) In mildew-infected barley and wheat leaves, P_i content is either unaffected by infection or it increases slightly at sporulation (Walters and Ayres, 1981; Zulu *et al.*, 1991a; Scholes *et al.*, 1992; Wright, 1992). Such results are markedly different from those observed in rust-infected barley (Ahmad *et al.*, 1982, 1984) and wheat (Bennet and Scott, 1971) leaves, where whole-leaf P_i at least doubled in comparison to uninfected leaves. The reasons for these differences are unknown but may reflect the more intimate association of rust haustoria with the mesophyll cells of the host.

(ii) If photosynthesis is restricted by the rate of end-product synthesis, the rate of photosynthesis can be increased by feeding P_i via the transpiration stream (Leegood and Furbank, 1986). When P_i was fed to rusted leaves of barley (Scholes and Farrar, 1986) and leek (Roberts and Walters, 1988), or to mildewed wheat (Zulu *et al.*, 1991a), the rate of photosynthesis was not increased.

If photosynthesis becomes inhibited as a result of a build-up (in the cytosol) of sucrose, or, in infected leaves, of glucose and fructose, it may be expected that hexose phosphates (hexose-P) will accumulate. In both mildewed leaves of barley and wheat, Fru6P and Glc6P increased as infection progressed (Scholes *et al.*, 1992). Their effect on the rate of sucrose synthesis and photosynthesis will depend upon their location – cytosolic or chloroplastic. Fru6P is located mainly in the chloroplast, whereas Glc6P may be located in the chloroplast or cytosol (this occurs because phosphogluco-

Table 7.2. *Maximal extractable activity of NADP-dependent glyceraldehyde-3-phosphate dehydrogenase and stromal fructose-1,6-bisphosphatase during steady state photosynthesis in wheat and barley leaves 5 days after inoculation with powdery mildew*

Enzyme	Enzyme activity (μmol mg^{-1} chlorophyll h^{-1})			
	Wheat		Barley	
	Uninfected	Infected	Uninfected	Infected
Stromal Fru1,6Pase	90.3 ± 2.8	63.8 ± 6.8	186.0 ± 16	100.6 ± 20
NADP-GAPDH	1191 ± 32	878 ± 120	1284 ± 223	964 ± 101

Data from Scholes *et al.* (1992) and Wright (1992).

isomerase is displaced from equilibrium in the chloroplast (Dietz, 1985)). The Glc6P: Fru6P ratio is therefore a rough indicator of the localization of these metabolites, a low ratio of XX (1.0) indicating localization in the stroma and a ratio of 3.0–4.0 localization in the cytosol. In infected leaves, the ratio of these metabolites (2.0) was identical to control leaves until late after infection (Scholes *et al.,* 1992). Thus, although the amount of hexose-P certainly increased in infected leaves, photosynthesis may not have been limited to any great extent by the availability of P_i. This is consistent with the above observations.

7.3.3 *What is the evidence for a direct effect of accumulation of carbohydrates on photosynthetic proteins?*

Studies of infected cereal leaves showed that photosynthesis was decreased because the Calvin cycle had been inhibited. Firstly, as mentioned previously, the inhibition of photosynthesis was greatest at saturating light and ambient CO_2 (Figure 7.5), when photosynthesis is predominantly controlled by the rate of carboxylation. Secondly, the activity of three enzymes of the Calvin cycle decreased markedly as infection progressed. Five days after inoculation of wheat leaves with mildew, the maximum activity of Rubisco was 50% of that in a control leaf (131 ± 26 compared to 270 ± 56 μmol mg^{-1} chlorophyll h^{-1}) and that of stromal Fru1,6Pase was reduced by approximately 30% in both mildewed leaves of wheat and barley (Table 7.2). NADP-glyceraldehyde phosphate dehydrogenase (GAPDH) was slightly (but not significantly) reduced (Table 7.2). Thirdly, changes in the ratios of specific metabolites were consistent with the decrease in enzyme activity. As infection progressed there were increases in the ratios of Ru1,5P_2 : PGA and triose-P : Ru1,5P_2, the first indicating a restriction in the carboxylation of CO_2 and the second a restriction in the regeneration of Ru1,5P_2 (Figure 7.6). The triose-P : Ru1,5P_2 ratio increased in mildewed leaves indicating that the reduction of PGA to triose-P was actually faster in these leaves (Figure 7.6).

The loss of activity and amount of photosynthetic proteins, particularly Rubisco, has been reported for a number of host–pathogen systems. In powdery mildew-

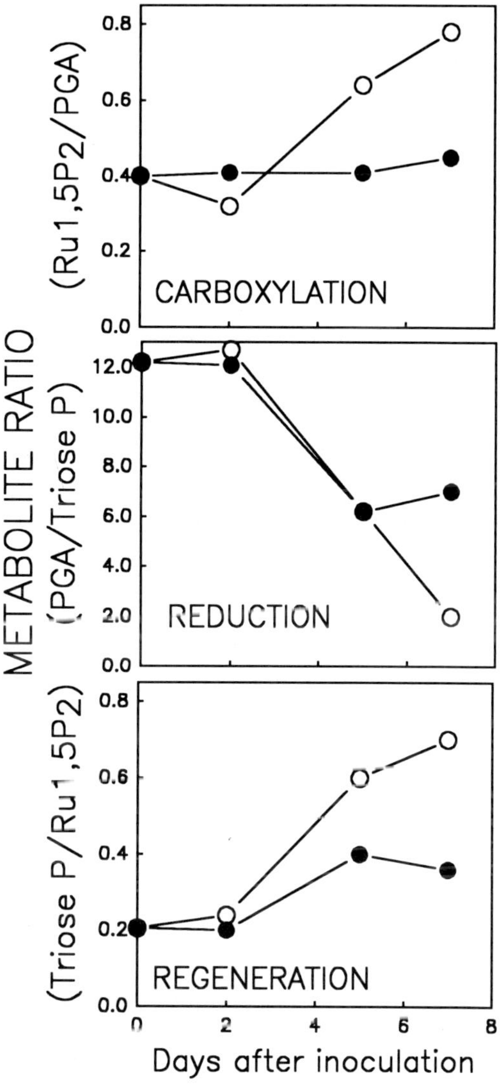

Figure 7.6. *Effect of powdery mildew on the metabolite ratios describing the partial reactions of the Calvin cycle, throughout infection; Ru1,5P$_2$ carboxylation (Ru1,5P$_2$:PGA); 3-PGA reduction (3-PGA:triose-P); Ru1,5P$_2$ regeneration (triose-P:Ru1,5P$_2$). Control (●) and infected (○) leaves. Results are the mean ± s.e. of four measurements. Modified from Scholes* et al. *(1992).*

infected sugar beet leaves a reduction in the amount and activity of Rubisco was correlated with a decrease in the rate of photosynthesis per unit leaf area and with a decline in the mesophyll conductance (Gordon and Duniway, 1982b). A similar reduction in the activity of this protein was observed in leaves of leek infected with *Puccinia allii* (Roberts and Walters, 1988) and of barley infected with powdery mildew (Walters and Ayres, 1984). The latter authors also observed a decrease in the activity of 3-phosphoglycerate kinase and NADP-GAPDH. Following infection of leaves of melon with *Colletotrichum lagenarium* (Roby *et al.,* 1988; Ranty *et al.,* 1987), potato with *Phytophthora infestans* (Kombrink and Hahlbrock, 1990) and barley with *Erysiphe graminis* (Higgins *et al.,* 1985), there was a progressive reduction in host mRNAs encoding both the large (LS) and small (SS) subunits of Rubisco. In the melon leaves infected with *C. lagenarium*, the two genes were differentially regulated during the course of infection and the breakdown of the subunits was controlled at several levels: transcriptional, post-transcriptional, translational and/or post-translational (Roby *et al.,* 1988).

7.3.4 *Regulation of electron transfer in cereal leaves infected with powdery mildew*

In Figure 7.1 it is suggested that electron transfer is down-regulated as a consequence of the decrease in the activity of the Calvin cycle, rather than as a primary effect of the pathogen. In both wheat and barley leaves infected with powdery mildew this hypothesis is supported by (i) no change in the initial slope of the light response curve of photosynthesis (and thus quantum yield) until late after infection (Scholes *et al.,* 1992; Wright, 1992) and (ii) analysis of chlorophyll fluorescence data. In the early stages of the development of powdery mildew on barley leaves (up to 4 dai) there was no evidence of a direct effect of the pathogen on PS II. Photochemical quenching (qP) (an indicator of the redox state of QA, the primary electron acceptor of PS II (Horton and Bowyer, 1990)), non-photochemical quenching (qN) (a measure of non-photochemical dissipation processes (Horton and Bowyer, 1990)) and ΦPS II (a measure of the quantum yield of PS II (Genty *et al.,* 1989)), were similar in infected and control leaves (Scholes *et al.,* 1990). As photosynthesis became inhibited there was an increased reduction of QA and an increase in qN, both indicative of a restriction in the thylakoid membranes. This is consistent with the reduced supply of PGA. It is probable that the gradual loss of chlorophyll was a consequence of the over-energization and over-reduction of the thylakoid membranes.

Some studies, however, have reported specific effects of pathogens on the partial reactions of electron transfer and/or on components of the thylakoid membranes. For example, Montalbini and Buchanan (1974) and Magyarosy *et al.* (1976) reported that the decreased rate of photosynthesis in chloroplasts isolated from broad bean leaves infected with *Uromyces viciae-fabae* and sugar beet leaves infected with *Erysiphe polygoni* resulted from a preferential inhibition of non-cyclic photophosphorylation. In addition, Magyarosy and Malkin (1978), again using powdery mildew-infected sugar beet leaves, suggested that the decline in the rate of photosynthesis could be attributed to a loss of the low and high potential forms of cytochrome b_{559}. In both these studies chloroplasts were isolated from leaves at a late stage of the infection cycle, therefore it is not possible to say whether the changes in electron transfer were the primary cause of the loss of photosynthetic activity, or merely a consequence of earlier events, as suggested above. Other investigations support the hypothesis that photosynthetic electron transfer was down-regulated in response to changes in metabolic demand rather than by direct effects of the pathogens. Neither Wynn (1963), working with chloroplasts isolated from rust-infected oat leaves, Ahmad *et al.* (1983) using barley leaves infected with brown rust, or Holloway *et al.* (1992) examining chloroplasts from barley leaves infected with powdery mildew, observed any reduction in non-cyclic electron transfer. The latter authors also demonstrated that the basis of expression of results was critical in interpreting the data. When results were expressed per unit fresh weight of leaf, there was a decline in electron flow from water through PS II and PS I to ferricyanide, through PS II and the plastoquinone pool only, and in cyclic electron flow around PS I in infected leaves. In addition, there was a reduction in cytochrome b_{559}, cytochrome *f* and P700, the reaction centre of PS I. However, when the data were expressed per unit of chlorophyll, there was no change in any of the partial reactions of electron transfer measured, or in components of the thylakoid membranes in rusted leaves when compared with controls. This clearly illustrates the heterogeneity in chloroplast populations within the infected tissue.

7.4 Conclusions

As a working hypothesis, it is proposed that the decline in the rate of photosynthesis in cereal leaves infected with powdery mildew occurs *primarily* as a result of an accumulation of carbohydrates following an increase in the activity of invertase in the infected leaf. Thus, the pathogen acts not by producing novel toxins or metabolites but by altering the source/sink relations of the leaf. In mildewed leaves, the changes in chlorophyll fluorescence parameters, amounts of carbohydrates and intermediates of the Calvin cycle and glycolysis, and the activities and amounts of enzymes of these pathways, exactly parallel those observed in the transgenic tobacco and *Arabidopsis* plants which overexpress yeast invertase in their cell walls (von Schaewen *et al.*, 1990; Stitt *et al.*, 1991) and leaves which had been fed glucose (Krapp *et al.*, 1991). I suggest that it is the increase in invertase activity in the infected leaves which initiates and maintains the changes in carbohydrate metabolism, yet many questions remain unanswered if the role of this enzyme in pathogenesis is to be fully understood. For example, how are the changes in invertase activity regulated? What is the signal transduction pathway that leads to the increase in activity?

Acknowledgements

I thank Professor D.H. Lewis for many helpful discussions and for critically reading the manuscript, and the Royal Society for a 1983 University Research Fellowship.

References

Ahmad, I., Owera, S.A., Farrar, J.F. and Whitbread, R. (1982) The distribution of five major nutrients in barley plants infected with brown rust. *Physiol. Plant Pathol.* **21,** 335–346.

Ahmad, I., Farrar, J.F. and Whitbread, R. (1983) Photosynthesis and chloroplast functioning in leaves of barley infected with brown rust. *Physiol. Plant Pathol.* **23**, 411–419.

Ahmad, I., Farrar, J.F. and Whitbread, R. (1984) Fluxes of phosphorus in leaves of barley infected with brown rust. *New Phytol.* **98,** 361–375.

Ayres, P.G. (1976) Patterns of stomatal behaviour, transpiration and CO_2 exchange in pea following infection by powdery mildew (*Erysiphe pisi*). *J. Exp. Bot.* **27**, 1196–1205.

Ayres, P.G. (1981) Effects of disease on plant water relations. In: *Effects of Disease on the Physiology of the Growing Plant* (ed. P.G. Ayres). Cambridge University Press, pp. 131–148.

Bagnall, D.J., King, K.W. and Farquhar, G.D. (1989) Temperature-dependent feedback inhibition of photosynthesis in peanut. *Planta* **175,** 348–354.

Bennet, J. and Scott, K.J. (1971) Inorganic polyphosphates in the wheat stem rust fungus and in rust infected wheat leaves. *Physiol. Plant Pathol.* **1,** 185–198.

Benhamou, N., Grenier, J. and Chrispeels, M.J. (1991) Accumualtion of β-fructosidase in the cell walls of tomato roots following infection by a fungal wilt pathogen. *Plant Physiol.* **97,** 739–750.

Besford, R.T. (1990) The greenhouse effect: acclimation of tomato plants growing in high CO_2, relative changes in Calvin cycle enzymes. *J. Plant Physiol.* **136,** 458–463.

Besford, R.T., Ludwig, L.J. and Withers, A.C. (1990) The greenhouse effect: acclimation of tomato plants growing in high CO_2, photosynthesis and ribulose-1,5-carboxylase protein. *J. Exp. Bot.* **41,** 925–931.

Billett, E.E., Billett, M.A. and Burnett, J.H. (1977) Stimulation of maize invertase activity following infection by *Ustilago maydis*. *Phytochem.* **16**, 1163–1166.

Brem, S., Rast, D.M. and Ruffner, H.-P. (1986) Partitioning of photosynthate in leaves of *Vitis vinifera* infected with *Uncinula necator* or *Plasmopora viticola*. *Physiol. Mol. Plant Pathol.* **29,** 285.

Callow, J.A., Long, D.E. and Lithgow, E.D. (1980) Multiple molecular forms of invertase in maize smut infections. *Physiol. Plant Pathol.* **16**, 93–107.

Daly, J.M. (1976) The carbon balance of diseased plants: changes in respiration, photosynthesis and translocation. In: *Encyclopedia of Plant Physiology,* volume 4 (eds R. Heitefuss and P.H. Williams). Springer, New York, pp. 450–479.

Dietz, K.-J. (1985) A possible rate-limiting function of chloroplast hexosemonophosphate isomerase in starch synthesis of leaves. *Biochim. Biophys. Acta* **839**, 240–248.

Dietz, K.J. and Foyer, C.H. (1986) The relationship between phosphate status and photosynthesis in leaves. Reversibility of the effects of phosphate deficiency on photosynthesis. *Planta* **167,** 376–381.

Doehlert, D. and Huber, S.C. (1983) Regulation of spinach leaf sucrose phosphate synthase by glucose-6-phosphate, inorganic phosphate and pH. *Plant Physiol.* **73**, 989–994.

Doehlert, D. and Huber, S.C. (1984) Phosphate inhibition of spinach leaf sucrose phosphate synthase as affected by glucose-6-phosphate and phosphoglucoisomerase. *Plant Physiol.* **76,** 250–253.

Donaldson, I.A. and Jorgenson, J.H. (1988) Barley powdery mildew 'invertase' is an α-glucosidase. *Carlsberg Res. Commun.* **53,** 421–430.

Farquhar, G.D. and Sharkey, T.D. (1982) Stomatal conductance and photosynthesis. *Ann. Rev. Plant Physiol.* **33,** 317–345.

Farrar, J.F. and Rayns, F.W. (1987) Respiration of leaves of barley infected with powdery mildew: increased engagement of the alternative oxidase. *New Phytol.* **107**, 119–125.

Farrar, J.F. and Lewis, D.H. (1987) Nutrient relations in biotrophic infections. In: *Fungal Infection of Plants* (eds. G.F. Pegg and P.G. Ayres). Cambridge University Press, Cambridge, pp. 92–132.

Foyer, C.H. and Spencer, C. (1986) The relationship between phosphate status and photosynthesis in leaves. *Planta* **167,** 369–375.

Foyer, C.H. (1987) The basis of source-sink regulation in leaves. *Plant Physiol. Biochem.* **25**, 649–657.

Foyer, C.H., Furbank, R., Harbinson, J. and Horton, P. (1990) The mechanisms contributing to photosynthetic control of electron transport by carbon assimilation in leaves. *Photosynth. Res.* **25**, 83–100.

Geiger, D.R. (1976) Effects of translocation and assimilate demand on photosynthesis. *Can. J. Bot.* **54,** 2337–2345.

Genty, B., Briantais, J.M. and Baker, N.R. (1989) The relationship between the quantum yield of photosynthetic electron transport and quenching of chlorophyll fluorescence. *Biochim. Biophys. Acta* **990**, 87–92.

Gerhardt, R., Stitt, M. and Heldt, H.W. (1987) Subcellular metabolite levels in spinach leaves. Regulation of sucrose synthesis during diurnal alterations in photosynthetic partitioning. *Plant Physiol.* **83**, 399–407.

Giaquinta, R.T. (1983) Phloem loading of sucrose. *Ann. Rev. Plant Physiol.* **34,** 347–387.

Gordon, T.R. and Duniway, J.M. (1982a) Photosynthesis in powdery mildewed sugar beet leaves. *Phytopathol.* **72**, 718–723.

Gordon, T.R. and Duniway, J.M. (1982b) Effects of powdery mildew infection on the efficiency of CO_2 fixation and light utilisation by sugar beet leaves. *Plant Physiol.* **69**, 139–142.

Greenland, A.J. and Lewis, D.H. (1981) The acid invertases of the developing third leaf of oat. II. Changes in the activities of soluble isoenzymes. *New Phytol.* **88**, 279–288.

Greenland, A.J. and Lewis, D.H. (1983) Changes in the activities and predominant molecular forms of acid invertase in oat leaves infected by crown rust. *Physiol. Plant Pathol.* **22**, 293–312.

Gwary, D.M. (1988) *Carbohydrate metabolism of barley infected with brown rust.* Ph.D. thesis, University of Wales.

Herold, A. (1980) Regulation of photosynthesis by sink activity – the missing link. *New Phytol.* **86,** 131–144.

Herold, A. and Lewis, D.H. (1977) Mannose and green plants: occurrence, physiology and metabolism, and use as a tool to study the role of orthophosphate. *New Phytol.* **79,** 1–40.

Herzog, B., Stitt, M. and Heldt, H.W. (1984) Control of photosynthetic sucrose synthesis by fructose-2,6-bisphosphate. III. Properties of the cytosolic fructose-1,6-bisphosphatase. *Plant Physiol.* **75**, 561–565.

Hewitt, H.G. and Ayres, P.G. (1975) Changes in CO_2 and water vapour exchange rates in leaves of *Quercus robur* infected by *Microsphaera alphitoides* (powdery mildew). *Physiol. Plant Pathol.* **7**, 127–137.

Higgins, C.M., Manners, J.M. and Scott, K.J. (1985) Decrease in three messenger RNA species coding for chloroplast proteins in leaves of barley infected with *Erysiphe graminis* f.sp. *hordei*. *Plant Physiol.* **78,** 891–894.

Holloway, P.J., Maclean, D.J. and Scott, K.J. (1992) Electron transport in thylakoids isolated from barley leaves infected by the powdery mildew fungus (*Erysiphe graminis* f. sp. *tritici*). *New Phytol.* **120,** 145–151.

Horton, P. and Bowyer, J.R. (1990) Chlorophyll fluorescence transients. In: *Methods in Plant Biochemistry,* volume 4 (Lipids, Membranes and Light Receptors) (eds J.R. Bowyer and J. Harwood). Academic Press, New York, pp. 259–296.

Huber, J.L.A., Huber, S.C. and Nielsen, T.H. (1989) Protein phosphorylation as a mechanism for regulation of spinach leaf sucrose phosphate synthase activity. *Arch. Biochem. Biophys.* **270,** 681–690.

Huber, S.C. and Huber, J.L. (1991) Regulation of maize leaf sucrose phosphate synthase by protein phosphorylation. *Plant Cell Physiol.* **32**, 319–326.

Klecan, A.L. and Buchanan, B.B. (1988) Powdery mildew infection alters fructose-2,6-bisphosphate content and sucrose to starch ratio in leaves of wheat plants actively forming grain. *Physiol. Mol. Plant Pathol.* **32,** 221–227.

Kombrink, E. and Hahlbrock, K. (1990) Rapid, systemic repression of the synthesis of ribulose 1,5-bisphosphate carboxylase small sub-unit mRNA in fungus-infected or elicitor-treated potato leaves. *Planta* **181,** 216–219.

Krapp, A., Quick, W.P. and Stitt, M. (1991) Ribulose-1,5-bisphosphate carboxylase-oxygenase, other Calvin cycle enzymes, and chlorophyll decrease when glucose is supplied to mature spinach leaves via the transpiration stream. *Planta* **186,** 58–69.

Krause, G.H. and Weis, E. (1991) Chlorophyll fluorescence and photosynthesis. The basics. *Ann. Rev. Plant Physiol. Plant Mol. Biol.* **42**, 313–349.

Krishnan, H.B. and Pueppke, S.G. (1988) Invertases from rust-infected wheat leaves. *J. Plant Physiol.* **133**, 336–339.

Leegood, R.C., Labate, C.A., Huber, S.C., Neuhaus, H.E. and Stitt, M. (1988) Phosphate sequestration by glycerol and its effects on photosynthetic carbon assimilation by leaves. *Planta* **176,** 117–126.

Leegood, R.C. and Furbank, R.T. (1986) Stimulation of photosynthesis by 2% oxygen at low temperatures is restored by phosphate. *Planta* **168,** 84–93.

Long, D.E., Fung, A.K., McGee, E.E.M., Cooke, R.C. and Lewis, D.H. (1975) The activity of invertase and its relevance to the accumulation of storage polysaccharides in leaves infected by biotrophic fungi. *New Phytol.* **74**, 173–182.

MacDonald, P.W. and Strobel, G.A. (1970) Adenosine diphosphate-glucose pyrophosphorylase control of starch accumulation in rust-infected wheat leaves. *Plant Physiol.* **46**, 126–135.

MacDonald, F.D. and Buchanan, B.B. (1990) The reductive pentose phosphate pathway and its regulation. In: *Plant Physiology, Biochemistry and Molecular Biology* (eds D.T. Dennis and D.H. Turpin). Longman, Singapore, pp. 239–252.

Magyarosy, A.C., Schürmann, P. and Buchanan, B.B. (1976) Effects of powdery mildew infection on photosynthesis by leaves and chloroplasts of sugar beet. *Plant Physiol.* **57**, 486–489.

Magyarosy, A.C. and Malkin, R. (1978) Effect of powdery mildew infection of sugar beet on the content of electron carriers in chloroplasts. *Physiol. Plant Pathol.* **13**, 183–188.

Manners, J.M. and Gay, J.L. (1983) The host-parasite interface and nutrient transfer in biotrophic parasitism. In: *Biochemical Plant Pathology* (ed. J.A. Callow). John Wiley and Sons, Chichester, pp. 163–195.

Mignucci, J.S. and Boyer, J.S. (1979) Inhibition of photosynthesis and transpiration in soybean infected with *Microsphaera diffusa*. *Phytopathol.* **69**, 227–230.

Mitchell, D.T. (1979) Carbon dioxide exchange by infected first leaf tissues susceptible to wheat stem rust. *Trans. Br. Mycol. Soc.* **72**, 63–68.

Montalbini, P. and Buchanan, B.B. (1974) Effect of rust infection on photophosphorylation by isolated chloroplasts. *Physiol. Plant Pathol.* **4**, 191–196.

Owera, S.A.P., Farrar, J.F. and Whitbread, R. (1981) Growth and photosynthesis in barley infected with brown rust. *Physiol. Plant Pathol.* **18**, 79–90.

Pegg, G.F. (1981) Growth regulators in the diseased plant. In: *Effects of Disease on the Physiology of the Growing Plant* (ed. P.G. Ayres). Cambridge University Press, Cambridge, pp. 47–84.

Preiss, J. (1982) Regulation of the synthesis and degradation of starch. *Ann. Rev. Plant Physiol.* **33**, 431–454.

Raggi, V. (1980) Correlation of CO_2 compensation point (Γ) with photosynthesis and respiration and CO_2-sensitive Γ in rust-affected bean leaves. *Physiol. Plant Pathol.* **16**, 19–24.

Ranty, B., Roby, D., Cavalie, G. and Esquerré-Tugayé, M.-T. (1987) Ribulose-1,5-bisphosphatase carboxylase/oxygenase expression in melon plants infected with *Colletotrichum lagenarium*. *Planta* **170,** 386–391.

Roberts, A.M. and Walters, D.R. (1988) Photosynthesis in discrete regions of leek infected with the rust, *Puccinia allii* Rud. *New Phytol.* **110**, 371–376.

Roby, D., Marco, Y. and Esquerré-Tugayé, M.T. (1988) Ribulose-1,5-bisphosphate carboxylase/oxygenase gene expression in melon plants infected with *Colletotrichum lagenarium*. Activity level and rate of synthesis of mRNAs coding for the large and small subunits. *Physiol. Mol. Plant Pathol.* **32**, 411–424.

Samborski, D.J. and Shaw, M. (1956) The physiology of host–parasite relations. II. The effect of *Puccinia graminis tritici* Eriks. and Henn. on the respiration of the first leaf of resistant and susceptible species of wheat. *Can. J. Bot.* **34**, 601–619.

Sawada, S., Hasegawa, Y., Kasai, M. and Sasaki, M. (1989) Photosynthetic electron transport and carbon metabolism during altered sink/source balance in single rooted soybean leaves. *Plant Cell Physiol.* **30,** 691–698.

Scholes, J.D. and Farrar, J.F. (1985) Photosynthesis and chloroplast functioning within individual pustules of *Uromyces muscari* on bluebell leaves. *Physiol Plant Pathol.* **27**, 387–400.

Scholes, J.D. and Farrar, J.F. (1986) Increased rates of photosynthesis in localised regions of a barley leaf infected with brown rust. *New Phytol.* **104**, 601–612.

Scholes, J.D., Lee, P., Horton, P. and Lewis, D.H. (1990) Photosynthetic metabolism in leaves infected with powdery mildew. In: *Current Research in Photosynthesis,* Volume 4 (ed. M. Baltscheffsky). Kluwer, Dordrecht, pp. 219–222.

Scholes, J.D., Lee, P.J., Horton, P. and Lewis, D.H. (1992) Invertase: the key to understanding changes in the photosynthetic and carbohydrate metabolism of barley leaves infected with powdery mildew (*Erysiphe graminis* f. sp. *hordei*). *New Phytol.* (in press).

von Schaewen, A., Stitt, M., Schmidt, R., Sonnewald, U. and Willmitzer, L. (1990) Expression of a yeast derived invertase in the cell wall of tobacco and *Arabidopsis* plants leads to inhibition of sucrose export, accumulation of carbohydrates and inhibition of photosynthesis and strongly influences growth and habit of transgenic tobacco plants. *EMBO J.* **9,** 3033–3044.

Scott, K.J. and Smillie, R.M. (1966) Metabolic regulation in diseased leaves. I. The respiratory rise in barley leaves infected with powdery mildew. *Plant Physiol.* **41**, 289–297.

Sheen, J. (1990) Metabolic repression of transcription in higher plants. *Plant Cell* **2**, 1027–1038.

Stitt, M. (1990a) The flux of carbon between the chloroplast and cytosol. In: *Plant Physiology, Biochemistry and Molecular Biology* (eds D.T. Dennis and D.H. Turpin). Longman, Singapore, pp. 309–326.

Stitt, M. (1990b) Fructose-2,6-bisphosphate as a regulatory molecule in plants. *Ann. Rev. Plant Physiol. Plant Mol. Biol.* **41**, 153–185.

Stitt, M., Gerhardt, R., Wilke, I. and Heldt, H.W. (1987) The contribution of fructose-2,6-bisphosphate to the regulation of sucrose synthesis during photosynthesis. *Physiol Plant* **69**, 377–386.

Stitt, M. and Quick, W.P. (1989) Photosynthetic carbon partitioning: its regulation and possibilities for manipulation. *Physiol. Plant* **77**, 633–641.

Stitt, M., von Schaewen, A. and Willmitzer, L. (1991) 'Sink' regulation of photosynthetic metabolism in transgenic tobacco plants expressing yeast invertase in their cell wall involves a down-regulation of the Calvin cycle and an up-regulation of glycolysis. *Planta* **183**, 40–50.

Sturm, A. and Chrispeels, M.J. (1990) cDNA cloning of carrot extracellular β-fructosidase and its expression in response to wounding and bacterial infection. *Plant Cell* **2**, 1107–1119.

Tetlow, I.J. (1990) *Photosynthetic carbon metabolism in plants infected with biotrophic fungi.* Ph.D. thesis, University of Wales.

Walker, D.A. and Sivak, M.N. (1986) Photosynthesis and phosphate: a cellular affair? *Trends Biochem. Sci.* **11**, 176–179.

Walters, D.A. and Ayres, P.G. (1981) Phosphate uptake and translocation by roots of powdery mildew infected barley. *Physiol. Plant Pathol.* **18,** 195–205.

Walters, D.R. and Ayres, P.G. (1984) Ribulose bisphosphate carboxylase protein and enzymes of CO_2 assimilation in barley infected by powdery mildew (*Erysiphe graminis hordei*). *Phytopath. Z. Bd.* **109**, 208–218.

Weil, M. and Rausch, T. (1990) Cell wall invertase in tobacco crown gall cells. *Plant Physiol.* **94**, 1575–1581.

Whipps, J.M. and Lewis, D.H. (1981) Patterns of translocation, storage and interconversion of carbohydrate. In: *Effects of Disease on the Physiology of the Growing Plant* (ed. P.G. Ayres). Cambridge University Press, Cambridge, pp. 47–84.

Whitney, S.H., Shaw, M. and Naylor, J.M. (1962) The physiology of host-parasite relations. Xii. A cytoplasmic study of the distribution of DNA and RNA in rust infected leaves. *Can. J. Bot.* **40,** 1433–1444.

Woodrow, I.E. and Berry, J.A. (1988) Enzymatic regulation of photosynthetic CO_2 fixation. *Ann. Rev. Plant Physiol. Mol. Biol.* **39,** 533–594.

Wong, S.C. (1979) Elevated atmospheric pressure of CO_2 and plant growth. Interaction of nitrogen and photosynthetic capacity in C_3 and C_4 plants. *Oecologia* **44,** 68–74.

Wright, D.P. (1992) *The effect of powdery mildew on the photosynthetic and carbohydrate metabolism of wheat leaves.* Ph.D. thesis, University of Sheffield.

Wynn, W.K. (1963) Photosynthetic phosphorylation by chloroplasts isolated from rust-infected oats. *Phytopathol.* **53**, 1376–1377.

Yelle, S.B., Beceson, R., Trudd, C. and Gosselin, A. (1989) Acclimation of two tomato species to high atmospheric CO_2. II Ribulose-1,5-bisphosphate carboxylase-oxygenase, and PEP carboxylase. *Plant Physiol.* **90,** 1465–1472.

Zulu, J.N., Farrar, J.F. and Whitbread, R. (1991a) Effects of phosphate supply on wheat seedlings infected with powdery mildew: carbohydrate metabolism of first leaves. *New Phytol.* **118**, 553–558.

Zulu, J.N., Farrar, J.F. and Whitbread, R. (1991b) Effects of phosphate supply on the phosphorus status, dry mass and photosynthesis of wheat infected with powdery mildew. *New Phytol.* **118**, 453–461.

8

Beyond photosynthesis: the translocation and respiration of diseased leaves

J.F. Farrar

8.1 Introduction

A diseased leaf typically photosynthesizes less than a healthy one, yet in addition to the normal uses for photosynthate there are the growth of the pathogen, host defence responses, and their associated respiration, as consumers of carbon skeletons. A smaller proportion of photosynthate will therefore be available for export in the phloem, resulting in a greatly reduced contribution by the diseased leaf to the carbon balance of the host. The fate of photosynthate in leaves infected with biotrophic fungi is considered in this chapter. What is known of control of translocation and respiration in healthy leaves is described as a prelude to the effects of infection. It is neither safe to generalize between diseases, nor possible to resolve any general question from work on a single disease, so unavoidably this account will be incomplete. The broad issues raised, and the mechanisms operating in the whole plant, should be as applicable to studies with pests as they are to infection by biotrophic pathogens.

8.2 Sucrose

Sucrose is the major product of photosynthesis, the major translocate, and the main substrate for metabolism in sinks in most plants (Figure 8.1). Not only does the reduced rate of photosynthesis in diseased leaves (Scholes, Chapter 7) reduce the rate of sucrose synthesis, but the extractable activity of the rate-limiting enzyme on the pathway to sucrose, sucrose phosphate synthase (SPS), is reduced in rust-infected barley, although it shows normal diel changes in activity (Tetlow and Farrar, 1992). We do not know if the reduced activity is a result of a reduced amount of SPS protein, but the activation state of SPS is dependent upon the rate of CO_2 assimilation in the short term (Battistelli *et al.*, 1991).

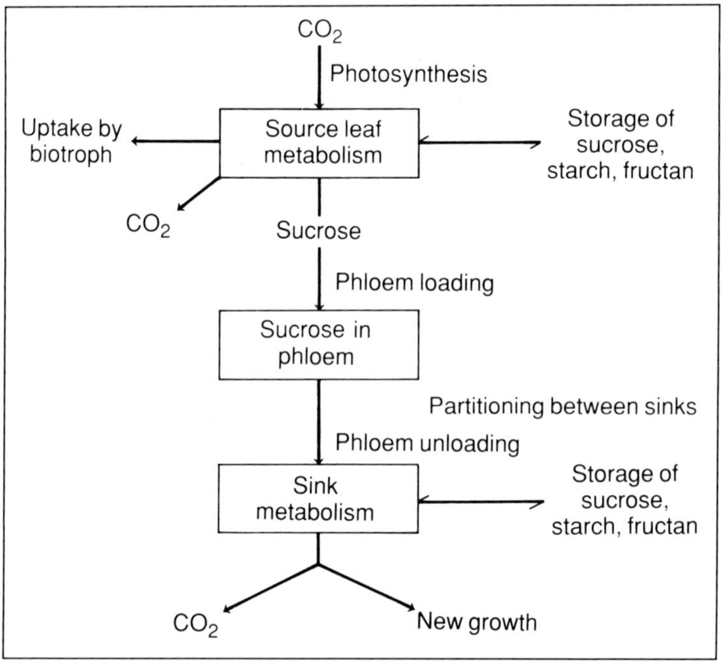

Figure 8.1. *Some partial processes of carbon partitioning and the central role of sucrose.*

Sucrose in healthy leaves is distributed between apoplast, cytosol and vacuole. As little as 20% of total sucrose may be cytosolic in grasses and cereals (Farrar and Farrar, 1986; Borland and Farrar, 1989), and up to 1% in the apoplast (Tetlow and Farrar, 1992). Whilst flux across the tonoplast is relatively slow, up to about 700 mg sucrose m^{-2} leaf h^{-1}, that across the plasmalemma is faster at about 1200 mg m^{-2} leaf h^{-1}, which is in excess of the maximum photosynthetic flux. The small apoplastic and cytosolic pools will therefore be in closer equilibrium with each other than will cytosolic and vacuolar pools. Export of sucrose is correlated with the size of the sucrose pool available for transport, considered to comprise the apoplastic plus the cytosolic pool (Farrar and Farrar, 1986). Autoradiographic evidence suggests that at least part of the transport pool is within vascular bundles of barley leaves, with the remainder in the mesophyll (Farrar and Farrar, unpublished). The flux through the transport pool is rapid, with a half-time of about 45 min, and cytosolic sucrose concentrations in barley are 20–150 mol m^{-3} (Farrar and Farrar, 1986).

Carbon from sucrose has three possible fates: storage in the vacuole as sucrose or fructan, export from the leaf in the phloem, or metabolism within the leaf, including respiration. In healthy barley leaves, about 40% of sucrose is stored temporarily in the vacuole before eventual export, about 5% is respired, and about 50% exported directly (Farrar and Farrar, 1985). The net storage of sucrose plus fructan by the end of a 12 h photoperiod is equivalent to approximately 5 h of photosynthetic sucrose production; most is mobilized during darkness to sustain translocation.

Fungi therefore invade a leaf with large pools of sucrose which are in communication with each other and turn over rapidly. Changes in sucrose flux can be manifested, in a very few hours, in altered pool sizes. The apoplast contains sucrose at a concentration of 1–10 mol m^{-3} which can readily be replaced from the cytosol across the plasmalemma.

The effects of fungi on sucrose content vary between species. Brown rust (*Puccinia hordei*) of barley reduces leaf sucrose content slightly (and hexoses greatly) at sporulation; the reduction is greater between than within pustules (Tetlow *et al.*, unpublished), a difference consistent with the higher rates of photosynthesis within pustules (Scholes and Farrar, 1986). The content of sucrose in the apoplast is lower, whether expressed as percentage of total or in absolute units, in rusted leaves, whilst the proportional distribution between cytosol and vacuole is unaltered (Tetlow and Farrar, unpublished). The flux into sucrose is reduced to about 60% of controls at sporulation, as a result of reduced photosynthesis and reduced activity of SPS, but there are still diel changes in sucrose content (Gwary and Farrar, unpublished). The fates of this sucrose form the subject of the remainder of this chapter.

8.3 Vacuoles

Especially in members of the Gramineae, vacuoles are important as stores of carbohydrate, especially sucrose and fructan. Carbohydrate within vacuoles is not readily available for metabolism, translocation, or uptake by a biotrophic fungus which is confined to the apoplast. There are profound changes in vacuolar carbohydrate metabolism in infected grasses and cereals. Classically, sucrose storage within vacuoles is negatively correlated with their content of acid invertase (ap Rees, 1974). The large increases in activity of acid invertase which follow infection by many biotrophic fungi (Farrar and Lewis, 1987) may be partly due to increased activity in vacuoles of the host (Tetlow and Farrar, 1992) which might be expected to reduce vacuolar storage of sucrose. A characteristic of sucrose-storing plants is their ability to synthesize fructan (Pollock, 1986). Fructan synthesis is stimulated, via *de novo* synthesis of sucrose–sucrose fructosyl transferase (SST), by high sucrose concentrations (Cairns and Pollock, 1989). Infected leaves commonly have massively increased fructan contents, the region within fungal pustules showing the increase in brown-rust infected barley (Gwary and Farrar, unpublished). Since pustule regions in this disease have higher rates of photosynthesis, and higher sucrose contents than non-pustule regions, sucrose accumulation may be the trigger for fructan synthesis. It is not clear if fructan accumulation is related to invertase activity, whether, for example, the invertase can function as an SST. Measurements of SST and of fructan hydrolase in infected leaves are badly needed.

Importantly, tonoplast properties may alter following infection. Barley infected with brown rust has a longer half-time than controls (31 h *vs.* 12 h) for loss of ^{14}C-sugars from vacuoles in experiments where leaf discs were cut from leaves which were previously fed $^{14}CO_2$ and were monitored for the washout of ^{14}C-sugars (Tetlow and Farrar, unpublished). Sucrose and fructan within vacuoles are therefore less readily available for translocation and metabolism than in healthy leaves. We do not

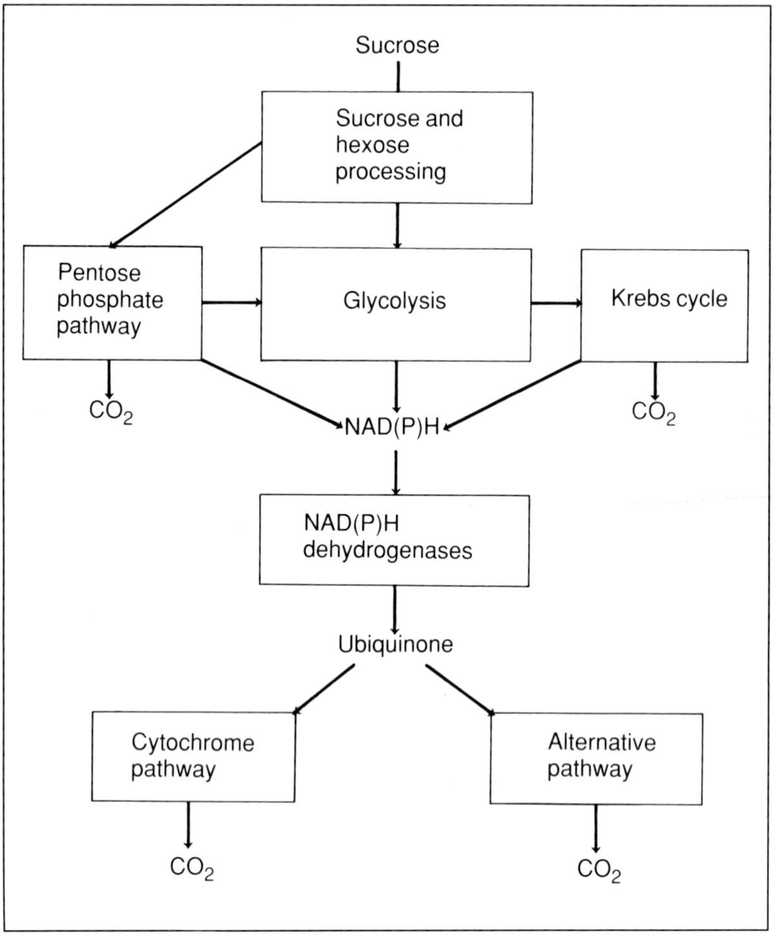

Figure 8.2. *Some component processes of plant respiration.*

understand the nature of changes to the tonoplast (which do not occur in oat infected with powdery mildew; Haigh and Farrar, unpublished), but they are likely to be solute-specific, as the efflux of $^{32}P_i$, $^{45}Ca^{2+}$, and $^{36}Cl^-$ across the tonoplast are not altered by rust infection (Ahmad *et al.*, 1984; Berryman and Farrar, unpublished).

8.4 Respiration

An outline of respiratory metabolism is given in Figure 8.2. Respiration results in the production of both energy and carbon skeletons, withdrawn from the respiratory network as substrates for biosynthesis. Healthy leaves respire both in the dark and, possibly at a reduced rate, in the light (Krömer and Heldt, 1991). The rate of respiration is typically insensitive to substrates such as sucrose, but is uncoupled by protonophores such as FCCP (Lambers, 1985), suggesting that oxygen uptake is

Table 8.1. *Rate of respiration of localized regions of fungally infected leaves: within and between fungal pustules*

Host and disease	Control leaf	Fungal pustules	Regions of infected leaves between pustules
Barley and brown rust[1]			
5 dai	2.9	8.5	4.5
8 dai	2.9	4.7	3.1
9 dai	2.3	4.0	1.9
Leek and rust[2]			
14 dai	0.48	1.59	0.31
Barley and powdery mildew[3]			
6 dai	0.29	0.43	0.13
12 dai	0.30	1.17	0.52
Wheat and stem rust[4]			
7–15 dai	0.59	1.12	0.70

Respiration as μmol $O_2\,m^{-2}$ leaf s^{-1}; dai, days after infection.
[1]Scholes and Farrar, 1986; [2]Roberts and Walters, 1988; [3]Bushnell and Allen, 1962; [4]Bushnell, 1970.

controlled not by substrate supply but by turnover of adenylates and thus demand for metabolic energy (Day *et al.*, 1985; Farrar and Lewis, unpublished; Pilbeam *et al.*, 1986). The cytochrome pathway carries the bulk of mitochondrial electron flux, and operates at 50–95% of its capacity, whilst the non-phosphorylating alternative pathway carries a smaller electron flux and is at just 10–50% of its capacity (Farrar and Williams, 1990). The ability of a mature leaf to produce ATP from mitochondrial respiration is therefore not greatly in excess of the normal rate of production: for barley, it is 150% of the normal, using the existing mitochondrial machinery, and for pea it is 125% (Farrar, unpublished). Mitochondria from green leaves can use glycine as a respiratory substrate, at a rate similar to that of substrates such as α-ketoglutarate and greater than that of malate (Douce, 1985), and a large capacity of mitochondrial electron transport is needed to cope with glycine delivered from photorespiration.

Leaves infected with biotrophic fungi commonly (but not invariably) have greatly increased rates of dark respiration (Table 8.1); rates of photorespiration are generally lower (Daly, 1976; Farrar and Lewis, 1987). An average mitochondrion will therefore sustain a much higher electron flux than controls, and a smaller proportion of this flux will come from the oxidation of glycine. Of course, there is no average mitochondrion. At least three sources can be distinguished: the fungus; the host, within pustules; and the host, between pustules.

Within pustules formed by rust fungi on leaves of barley (Scholes and Farrar, 1986), bluebell (Scholes and Farrar, 1985), leek (Roberts and Walters, 1988) and other hosts, respiration is substantially higher than in surrounding, uninfected tissue (Table 8.1). Since separating fungus and host from rusted leaves is not yet possible, we cannot attribute the increase in respiration to either host or fungus. However it is likely that both contribute to it. The fungus must support its own growth and metabolism, and a rough calculation suggests that all of the increase in respiration rate of a barley leaf infected with brown rust could be ascribed to the fungus (Kneale and Farrar, 1985). But since uninfected regions around pustules also sustain increased respiration, the host is clearly contributing to the overall increase in respiration in wheat stem rust (Bushnell, 1970), barley brown rust (Scholes and Farrar, 1986) and barley powdery mildew (Bushnell and Allen, 1962). With powdery mildew infections, it is possible simply to wipe most fungal biomass from the surface of the leaf. The leaf, plus haustoria in the epidermis, commonly respires at a rate little short of that before removing the superficial mycelium, and so most of the increased respiration has been attributed to the host (Daly, 1976). Host protoplasts isolated from leaves of barley infected with powdery mildew respire faster than those from healthy leaves (McAinsh *et al.*, 1989).

The rest of this discussion must be restricted, unsatisfactorily, to whole leaves and thus 'average' mitochondria. The rate of respiration commonly increases 3–4 fold after infection. Since this increase is more than the capacity of the mitochondria of control leaves (capacity being when both cytochrome and alternative pathways are operating at maximal rates), increases in the amount of respiratory machinery must have occurred. Neither direct measurements of mitochondrial machinery (fumarase, cytochrome C oxidase), nor careful use of inhibitors to assess capacity of the electron transport pathways, are available to support this statement for diseased leaves. However, comparisons of the uncoupled rate of respiration for barley leaves with the rate when infected with brown rust or powdery mildew support the idea, and suggest increases occur in the capacity of the cytochrome pathways (Lewis *et al.*, unpublished; Farrar and Rayns, 1987). For an isoline of wheat susceptible to stem rust, the capacity of the alternative pathway increased by 75% whilst the activity of the cytochrome pathway increased 3-fold (Haigh *et al.*, unpublished). Further, both alternative and cytochrome pathways are engaged in those diseases so far investigated (Figure 8.3; Farrar and Rayns, 1987; Haigh *et al.*, 1991; Gwary and Farrar, unpublished; Haigh *et al.*, unpublished; McAinsh *et al.*, 1989, 1991). For the cytochrome pathway, we can be reasonably sure that its increased engagement is due to increased demands for ATP and a faster turnover of adenylates. Further, the increase in amount of cytochrome pathway machinery that has been deduced is compatible with a hypothesis developed from work in roots. This proposes that increased flux through the cytochrome pathway (increased as a result of sucrose stimulation of energy-consuming processes) may cause greater reduction of the components of the chain, which could in turn trigger the synthesis of more machinery (Farrar and Williams, 1990), possibly following protein phosphorylation, by analogy with chloroplast enzymes (Scheibe, 1991). The control of the alternative pathway is not understood, nor is its role, and so it is far harder to interpret its response to infection.

Respiration has two major roles: the provision of carbon skeletons for biosynthesis, and the production of NAD(P)H and ATP. The increase in respiration rate of diseased leaves is not matched by a proportional increase in ATP production, since the non-phosphorylating alternative pathway accounts for an increased part of the total electron flux, but the flux of carbon compounds will increase with either pathway.

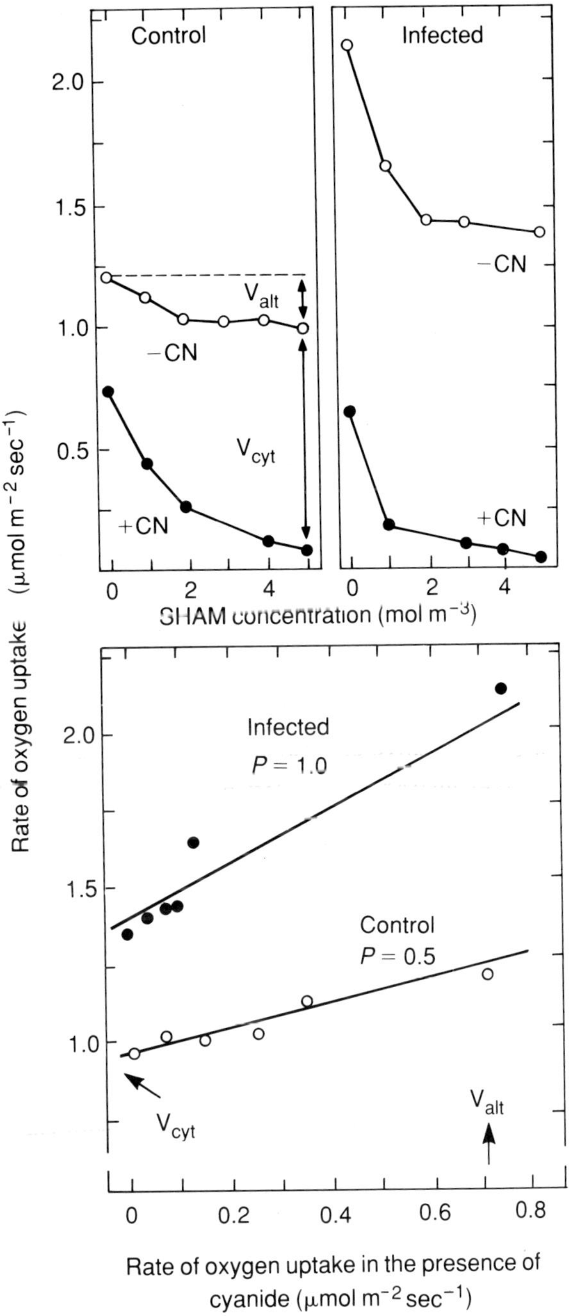

***Figure 8.3**. Respiration in leaves of barley (*cv. *Marris Otter) infected with powdery mildew. The respiration rate of control and infected leaves 9 dai, showing: (○) rate in the presence of increasing concentrations of salicylhydroxamic acid (SHAM), and (●) rate in the presence of 0.3 mol m^{-3} KCN and increasing concentrations of SHAM. Below, these data redrawn as a Bahr–Bonner plot from which the activity of the cytochrome pathway (V_{cyt}, intercept on *y*-*axis), engagement of the alternative pathway (P, slope of line), and capacity of the alternative pathway (V_{alt}, maximum value along *x*-*axis) can be determined; ○, control; ●, rusted.*

8.5 Respiratory carbon metabolism

There has been little work on respiratory carbon metabolism of infected leaves since the reviews of Daly (1976) and Chakravorty and Scott (1982). The basic conclusion that there is an increase in the flux through the pentose phosphate pathway relative to that through glycolysis still stands. Recent advances in the enzymology and control of glycolysis in healthy plants make it time to re-examine diseased plants. Glycolysis has long been believed to be controlled at phosphofructokinase (PFK), a conclusion generally supported by application of control theory. An alternative route from fructose 6-phosphate (Fru6P) to fructose 1,6-bisphosphate (Fru1,6P_2) is now characterized, using the pyrophosphate-dependent enzyme PFP (pyrophosphate:fructose-6-phosphate phosphotransferase). Control at PFK/PFP is partially exercised by the regulatory sugar fructose 2,6-bisphosphate (Fru2,6P_2), with a high concentration of Fru2,6P_2 potentially increasing glycolytic flux by activating PFP (Huber, 1986; Sung *et al.*, 1988). High concentration of Fru2,6P_2 may be the result of low 3-phosphoglycerate (PGA) and high inorganic phosphate (P_i) concentrations (Huber, 1986). Fru2,6P_2 also inhibits 6-phosphogluconate dehydrogenase, an enzyme of the pentose phosphate pathway, and activates UDPGlc-phosphorylase, which may be involved in sucrose degradation (Huber, 1986). Overall, high Fru2,6P_2 concentrations favour glycolysis (Stitt, 1987).

Leaves of wheat infected with stem rust have a reduced content of Fru2,6P_2 (Streuter *et al.*, 1989). If heterogeneity across the leaf is ignored, there will be a lower concentration of Fru2,6P_2 in host cells and, other things being equal, a lower rate of glycolysis. The lower concentration could relieve inhibition of 6-phosphogluconate dehydrogenase and so increase carbon flux through the pentose phosphate pathway. Since the pentose phosphate pathway provides a potential bypass of the PFK/PFP reactions of glycolysis, by partially degrading glucose 6-phosphate (Glc6P) to triose phosphate (triose-P) which can re-enter glycolysis, there is a consistency between Fru2,6P_2 and qualitative changes in respiratory carbon metabolism in rusted barley leaves. Much further work is needed to establish whether this story has general, or indeed any, validity.

In young wheat infected with powdery mildew, however, Fru2,6P_2 content increased on infection when the whole mildewed leaf was assayed (Klecan *et al.*, 1986), but the increase was apparent only long after infection, in leaves of mature wheat producing seed (Klecan and Buchanan, 1988). For oat infected with powdery mildew, the phosphate status of the plant affects foliar Fru2,6P_2 concentrations more than does infection, plants grown P_i-free having less Fru2,6P_2 than those grown on 3 mol m^{-3} P_i (Gunn *et al.*, unpublished). Since Fru2,6P_2 is involved in photosynthetic as well as respiratory metabolism, and changes in its concentration have not been localized within infected leaves (to within or between fungally invaded areas, or to different cell types, or indeed the fungus itself), it is premature to seek a firm role for Fru2,6P_2 in affecting the changed rate of respiration caused by infection. Indeed, a role in control of respiration in healthy plants is still disputed (Stitt, 1987).

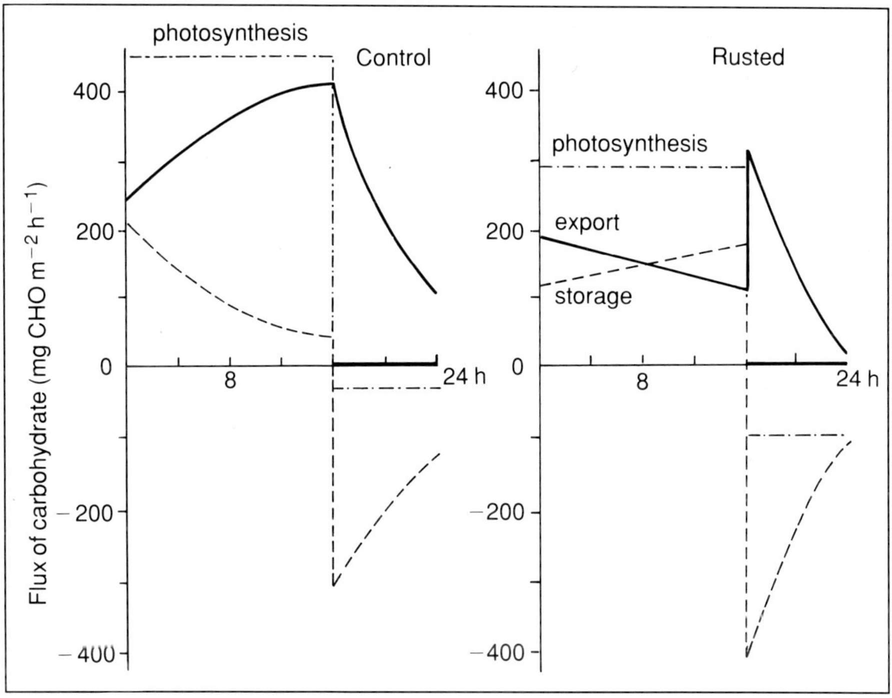

***Figure 8.4**. Diel carbon balance of leaves of barley, healthy or 8 days after infection with brown rust. Net CO_2 exchange (–·–·–) was measured continuously and the rate of carbohydrate storage or mobilization (sucrose, fructan and starch; ----) calculated after serial harvests and measurements of carbohydrate content. Rate of translocation (——) calculated as difference between gas exchange and storage. From data of Gwary and Farrar (unpublished).*

8.6 Translocation

Assimilate that is not stored, respired or metabolized is available for translocation. Because a diseased leaf typically has a lowered rate of photosynthesis, an additional sink for assimilate in the form of the pathogen, and an increased rate of respiration, less translocation from it would be expected, and this simple consequence of mass balance has been observed in many diseases (Figure 8.4; Whipps and Lewis, 1981; Farrar and Lewis, 1987). It certainly seems that translocation has lower priority as a fate for sucrose than metabolism by the host and uptake by the fungus. There have been substantial increases in our understanding of the partial processes of translocation over the last few years in healthy plants (Baker and Milburn, 1989; Wardlaw, 1990; Bonnemain *et al.*, 1991) and it is time these were applied to infected leaves.

Transport of assimilate across leaves from mesophyll to conducting tissue probably occurs symplastically, with diffusion as the driving force, possibly assisted by cyclosis. Sugars leaking into the apoplast can be actively reloaded into the symplast (Maynard and Lucas, 1982). Particular control of transport may be exercised by the parenchyma-

Table 8.2. *Apoplast in first leaves of barley infected with brown rust: volume, pH, sugar concentration and activity of extracellular invertase, at sporulation*

	Control	Rusted
Volume of apoplast ($cm^3\,m^{-2}$ leaf)	12	17
Tissue volume (%)	5.8	8.1
Concentration ($mol\,m^{-3}$)		
Sucrose	1.7	0.7
Glucose	1.4	0.5
Fructose	1.3	0.8
pH	6.63	7.25
Activity of extracellular invertase (nmol glucose $m^{-2}\,s^{-1}$) with 5 $mol\,m^{-3}$ sucrose as substrate	47	88

Data of Tetlow and Farrar (unpublished).

tous bundle sheath of grasses (Williams *et al.*, 1989) or the paraveinal mesophyll of legumes (Franceschi and Giaquinta, 1983), and it may be no coincidence that brown-rust of barley extends runner hyphae down the surface of the parenchymatous bundle sheath, sending haustoria into each of its cells (Scholes and Farrar, unpublished). Real controversy starts over the method by which sucrose enters the phloem. The paradigm has been that sucrose is unloaded into the apoplast around the sieve-element:companion-cell (se : cc) complex and is then loaded, by proton symport, into the se : cc complex (Delrot, 1989). Alternatively, there may be some species in which the se : cc complex is loaded partly or entirely symplastically through plasmodesmatal connections from surrounding cells (van Bel, 1987), and barley may be one such species (Farrar *et al.*, 1992).

A pathogenic fungus growing in a leaf must tap into the flux of assimilate between chloroplast and se : cc complex. Few biotrophic fungi invade the vascular tissue (an exception is *Puccinia poarum* on *Tussilago farfara*; Al-Khaesraji *et al.*, 1980), for example, powdery mildews are confined to epidermal cells and in all cases either the general or a specialized apoplast separates host and fungus (Manners and Gay, 1983). For powdery mildews the question becomes, how is assimilate directed not to phloem but to cells containing haustoria? For rusts, the same question can be asked but because, in addition, intercellular hyphae may take up nutrients (Kneale and Farrar, 1985), a second question obtains: how may assimilate be delivered to hyphae in the apoplast? Manners and Gay (1983) have addressed the first question and shown high activity of membrane-bound ATPases at the plasmalemma of the fungus or of the host cell enclosing it. A combination of the normal leakiness of mesophyll cells, and preferential localization of apoplastic retrieval mechanisms may be sufficient for local redirection of sugars. Transport at the host–parasite interface is discussed by Smith and Smith (1990).

The second question has two complementary answers. First, sugars continually enter the apoplast from mesophyll cells, which are leaky (Farrar and Farrar, 1986; Madore and Lucas, 1987). They achieve mean apoplastic concentrations of 1.7 and

2.7 mol m^{-3} for sucrose and hexose, respectively, in healthy barley leaves (Table 8.2; Tetlow and Farrar, unpublished). Since a typical K_m for sugar transport is 5 mol m^{-3} (Komor, 1982; Rheinhold and Kaplan, 1984) the concentrations are sufficient to permit rapid active uptake. The second answer is that conditions in the apoplast of diseased leaves may differ in a way that favours fungal acquisition of sugars. The pH of the apoplast of leaves of barley infected with brown rust is higher than in healthy leaves (Tetlow and Farrar, unpublished) and an increased activity of cell-wall invertase is often seen (Table 8.2; Farrar and Lewis, 1987). The increased pH means that the proton gradient for driving sucrose reloading from the apoplast has been reduced; consequently, the ratio of hexoses to sucrose rises. In barley infected by brown rust, apoplastic sucrose concentration falls to 41%, and that of hexose to 52%, of control values (Table 8.2). Overall there is a more favourable environment for uptake of hexose by the hexose carriers of the fungal hyphae.

8.7 Export in the phloem

The flux of sucrose from mesophyll cells to phloem will be reduced in tandem with flux in the phloem itself. Currently there is no need to look beyond the reduced availability of carbon in biotrophic infections to explain the reduced flux, but a closer examination of the partial processes may reveal other mechanisms by which pathogens act. Reduced availability of sucrose is the simplest of explanations, and it is supported by analysis of export from infected leaves fed $^{14}CO_2$. After a pulse of $^{14}CO_2$ lasting typically 20 min, pools of transport sucrose and starch are heavily labelled, while vacuolar sucrose is rather less so due to its long half-time. If a Geiger–Müller tube is attached to the leaf, the loss of ^{14}C from the leaf can be followed non-destructively (Figure 8.5). A simple calculation shows that respiratory loss of ^{14}C is negligible, so the curve describes export in the phloem. Typically, the efflux curve has three components: in the first $1\frac{1}{2}$–2 h, loss is largely from the transport pool of readily translocated sucrose; from 2 h to the time of darkening, loss is from sucrose stored in the vacuole; and during darkness starch is mobilized in addition to continuing loss of vacuolar sucrose (Farrar and Farrar, 1985, 1986). In some species, starch may turn over in the light (Farrar, 1991). Efflux curves produced in the light can be analysed as a two-compartment model in which the effects of pool sizes can be separated from those of rate constants. Rate constants describe the rate at which each unit of material in a pool is lost from it, and will reflect underlying mechanisms.

In barley infected with brown rust, translocation at sporulation is reduced greatly, but the rate constant for transfer of sucrose from the transport pool out of the leaf is quite unaffected (Figure 8.5; Table 8.3; Owera *et al.*, 1983; Tetlow and Farrar, unpublished), suggesting strongly that the mechanism of phloem loading is functioning perfectly well in the infected leaf. It is also unaffected in oats infected with powdery mildew (Haigh and Farrar, unpublished), where there is a small but significant reduction in translocation from infected first and fifth leaves. By careful screening of selected areas of the leaf, it is possible to show that the pustules and between-pustule regions of a brown-rust infected barley leaf behave quite differently, with the pustules showing net import of ^{14}C and the export from between-pustule regions being more

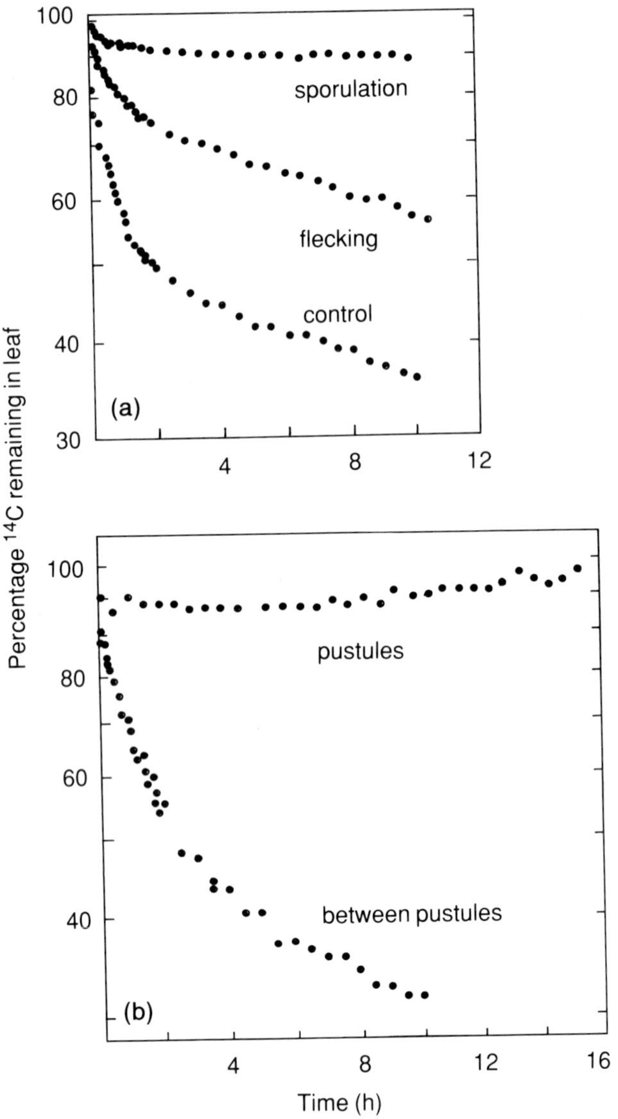

***Figure 8.5**. Translocation of carbon from leaves of barley infected with brown rust. Following feeding $^{14}CO_2$ to the leaf for 20 min, the leaf was attached to a Geiger–Müller tube and its ^{14}C content monitored continuously. (a) Efflux of ^{14}C from control leaves; infected leaves at flecking; and at sporulation; (b) efflux of ^{14}C from pustules and uninfected regions between pustules, on a sporulating leaf. Unpublished data of Tetlow and Farrar.*

Table 8.3. *Translocation from, and compartmentation in, leaves of barley infected with brown rust*

	Control	Rusted
Rate of translocation (mg CHO m^{-2} leaf h^{-1})	580	210
Size of transport pool ($g\,m^{-2}$)	0.33	0.12
Size of vacuolar pool ($g\,m^{-2}$)	1.59	63.6
Half-time of transport pool (h)	0.23	0.30
Half-time of vacuolar pool (h)	3.43	693.0

Unpublished data of Tetlow and Farrar.

like that of healthy leaves (Figure 8.5). Overall, carbon transfer through the transport pool from sucrose stored in the vacuole is reduced substantially by infection, as discussed earlier and shown by the reduced slow phase of ^{14}C efflux and the slow turnover of the vacuolar pool (Table 8.3; Figure 8.5).

More care must be taken when interpreting the behaviour of the 'vacuolar' pool, since the increased retention and metabolism of sugars in the infected leaf will appear to give a larger pool which turns over more slowly; however, the flux into structural material is not great enough to account for all of the changes to the vacuolar pool (Tetlow and Farrar, unpublished).

8.8 Partitioning of photoassimilate

The partitioning of assimilate between organs on a plant is frequently altered by pathogens. Shoot : root ratio (S : R) often increases, for example, so that a higher proportion of the assimilate exported from a leaf now goes to the shoot (Walters, 1985). The control of carbon partitioning between organs is not understood in healthy plants, although there are some recent advances. The consensus view is that phloem operates by turgor-driven bulk flow (the Münch hypothesis), so that flux J (mol s^{-1}) is given by

$$J = A \Delta P k c,$$

where A is cross-sectional area of sieve tubes (m^2), ΔP the pressure difference between source and sink (MPa), k the conductance of each unit cross-sectional area of phloem (m $MPa^{-1}\,s^{-1}$), and c solute concentration (mol m^{-3}). Velocity, v (m s^{-1}) = $\Delta P k$.

From the Hagen–Poiseuille equation:

$$k = r^2/8\eta l$$

where r and l are radius and length of sieve tubes or pores (m), and η is viscosity.

Thus a suite of variables could contribute to a change in flux down a given region of phloem. Between-organ partitioning, such as determines S:R, will then depend on the ratios of these variables. The determination of ΔP, the driving force, is particularly interesting. In both source and sink the turgor pressure P_{ph} in phloem is given by:

$$P_{ph} = \Psi_{ph} - \Psi_{apo}$$

where Ψ is solute (osmotic) potential in phloem (ph) and the apoplast (apo) surrounding it, respectively, and the status of source and sink must be communicated to phloem by altering Ψ_{ph} or Ψ_{apo} (Patrick, 1991; Williams *et al.*, 1991).

We do not yet understand the water relations of diseased plants in terms which enable us to address the control of carbon partitioning. But if we apply the two equations given at the start of this section, it is immediately clear that several parameters may be affected. With source leaves of barley infected with brown rust, solute potential is increased, there is a greater concentration of osmolytes, and in the same leaves the turgor pressure of epidermal cells is decreased. The osmolyte concentration in the xylem sap is lower than in controls, but there is no information on the phloem (Berryman *et al.*, 1991a, b, c).

We do know that the flux of C, N, P and K out of rusted or mildewed barley leaves is substantially reduced (Ahmad *et al.*, 1982, 1984; Murray and Ayres, 1986). As the flux of each of these is reduced, it is likely that total volume of solution moving down the phloem is lower; as their fluxes are differently reduced, it is probable that loading too is affected. For only very few biotrophic infections is there any suggestion that the pathogen invades vascular tissue (see above) reducing its functional area A or reducing k. For only one disease, barley infected with brown rust, is a measurement of speed of phloem transport available, and it increases slightly (Figure 8.6; Gwary and Farrar, unpublished). ΔP and c are left as the determinants of flux in the phloem that this disease may affect, and they are interrelated in that sucrose, whose concentration is defined by c, is the main generator of solute potential with the phloem, and thus of P in source leaf phloem. It seems likely that concentrations of sucrose in the phloem of rusted barley leaves are reduced, probably as a result of reduced loading. Additionally, apoplastic concentrations of osmolytes may be higher (Farrar and Lewis, 1987; Zulu *et al.*, 1991b) which would reduce P in source leaf phloem further; difference in P between source and sink would then be reduced, and thus so would carbon flux. The outcome of such changes in phloem transport will be considered next.

In healthy plants, S : R changes during development and thus with plant size. When a treatment that retards or accelerates growth is applied, that treatment may appear to alter S : R ratio whilst simply affecting growth. In order to show that development itself is indeed affected by treatment, it is necessary to apply the allometric formula:

$$S = bR^k$$

(which can be linearized to $\ln S = \ln b + k \ln R$) where S and R are shoot and root weights, k the allometric constant and b is a constant reflecting initial conditions. (See Paul, Chapter 9, for further discussion of allometric relationships).

Oat infected with powdery mildew shows significantly increased S : R, but k is not systematically altered when the infected oats are grown at 0 or 3 mol $P_i\,m^{-3}$ (Table 8.4; Gunn and Farrar, unpublished). Infection of barley with powdery mildew causes an increase in the value of k, and thus a real developmental switch towards increased growth of shoot relative to the root, as is clear from reanalysis of the data of Walters and Ayres (1981) and Last (1962) (Table 8.4; Figure 8.7). Although development of uninfected barley is such that S : R is continually rising ($k > 1$), it is clear that infection results in an increased S : R with a quite different cause.

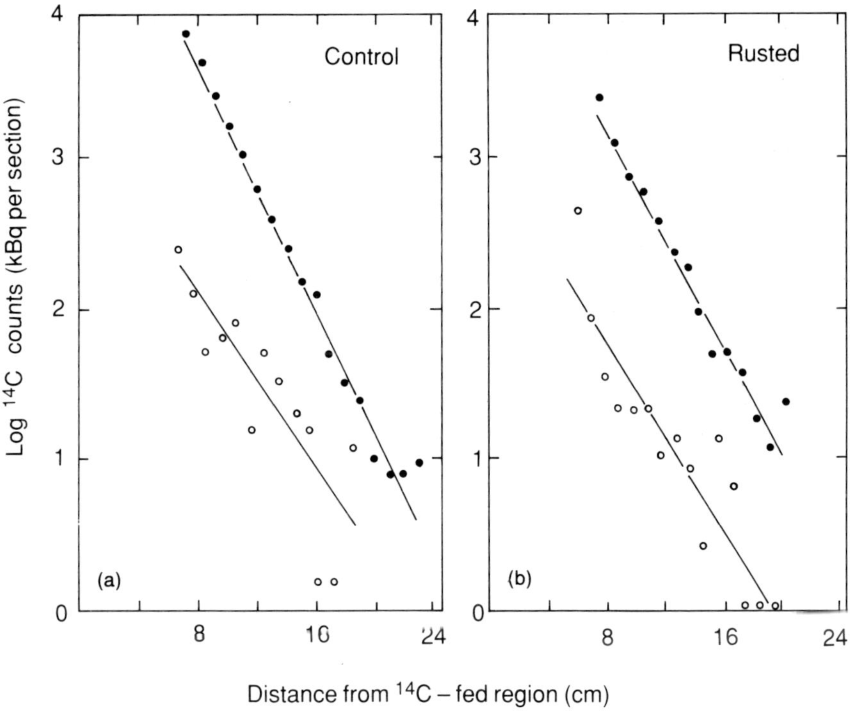

Figure 8.6. *Speed of phloem translocation in barley infected with brown rust. Tips of leaves were supplied with $^{14}CO_2$ for 10 min and the plants then either (○) harvested immediately, or (●) chased in air containing unlabelled CO_2 for 10 min before harvesting. Leaf blade (below the fed area), leaf sheath and root were rapidly cut into 1 cm portions and the radioactivity in them counted to obtain a spatial profile. Speed is calculated as the displacement between the two lines. (a) Control; (b) rusted (day of sporulation). Data of Gwary and Farrar (unpublished).*

Numerous studies with ^{14}C have followed the distribution of photosynthetically fixed carbon (Whipps and Lewis, 1981), but these may lack the sensitivity of studies examining the allometric coefficient. Percentage distribution of isotope, the manner in which data are often recorded, takes no account of dilution of isotopes into carbohydrate pools altered by infection, and calculation of flux from ^{14}C measurements is by no means straightforward. Walters and Ayres (1982) demonstrate changes in carbohydrate pools of roots of barley infected with powdery mildew, and changes in leaves have been mentioned above. Two other measures of the extent to which roots are affected may also be useful-root respiration rate, which is linearly related to growth rate (Farrar and Williams, 1990), and the mitotic index of root tips.

The root tips of barley infected with powdery mildew have a substantially reduced mitotic index, the reduction beginning when changes in carbohydrate metabolism could be detected in the root tips (Walters and Ayres, 1982). Since completion of the cell cycle in root tips is dependent on a continued supply of sucrose (van't Hof and Kovacs, 1972) a reduced mitotic index may indeed be a consequence of reduced

Table 8.4. *Shoot:root relations of plants infected with biotrophic fungi: application of the allometric relationship* $S = bR^K$ *(where* S *and* R *are shoot and root dry weights,* b *constant, and* K *the allometric constant)*

Host and disease	*b*		*K*	
	Control	Infected	Control	Infected
Oat and powdery mildew[1]				
Grown on 0 mol m^{-3} PO_4^{2-}	0.64	1.49	1.15	0.96
Grown on 3 mol m^{-3} PO_4^{2-}	1.30	1.30	1.09	1.11
Barley and powdery mildew[2]	0.92	0.82	1.05	1.12
Barley and powdery mildew[3]	1.11	0.83	1.24	1.51

[1]Gunn and Farrar, unpublished; [2]Walters and Ayres, 1981; [3]Last, 1962. Data from the last two references have been used to calculate *b* and *K*.

transport of assimilate to roots, with obvious implications for future growth of the roots. Respiration rate of roots of wheat is not affected when the leaves are infected with powdery mildew (Zulu *et al.*, 1991a).

Why should assimilate supply to the roots be reduced more than that to the developing shoot? It may be that a typical plant is organized so that the lower the sucrose flux from the leaves, the smaller the proportion of sucrose that goes to the root. Not only would this provide an explanation for some programmed responses of S : R to environmental change (Wilson, 1989) but it is also consistent with some preliminary experiments on barley (Minchin *et al.*, unpublished). It would happen if the root works at a rate nearer to its capacity than does the shoot: then, any fall in flux from the source would be accompanied by a change in partitioning in favour of the

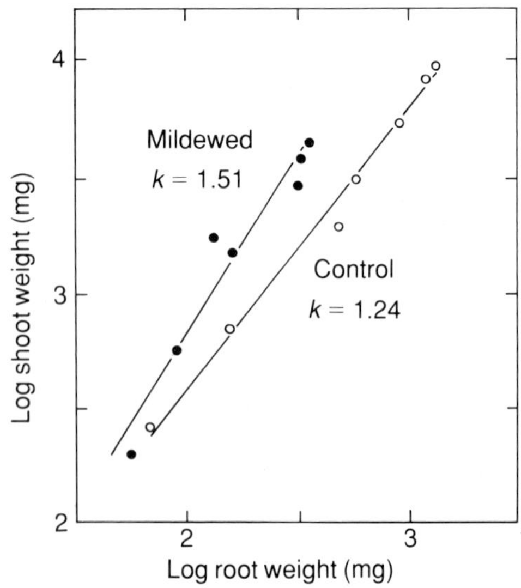

Figure 8.7. *Allometric growth of control (○) and mildewed (●) barley. Recalculated from data of Last (1962). For values of constants, see Table 8.4.*

shoot, a conclusion supported by a model of phloem transport (Minchin *et al.*, unpublished). Further, since meristems in the shoot are usually closer to source leaves than are root meristems, the turgor gradient would favour transport to the shoot over that to the root.

It is common for S : R to increase in infected plants (Walters, 1985). This increase has been ascribed to a functional equilibrium between shoot and root, where the product of shoot mass and activity of each unit of shoot is in proportion to an identical product for the root:

$$SA_S = KRA_R$$

where A_S and A_R are activity per unit weight of shoot and root, and K is a constant (Walters, 1985; Wilson, 1989). Whilst it is perfectly possible that it is a functional equilibrium that determines the outcome of partitioning in diseased plants, the simple consequence of a reduced flux from source leaves may account for altered S : R (and may, of course, be part of the mechanism by which a functional equilibrium is achieved).

Phloem unloading has not been explored in detail in diseased plants. It is clear that, in brown rust-infected barley leaves, carbon exported from the tip region can be unloaded and enter pustules further down the leaf. The intimate association between pustules and the vascular system may aid this directed export: not only do runner hyphae extend along the surface of the parenchymatous bundle sheath (Scholes and Farrar, unpublished) but pustules seem to be centred on veins rather than mesophyll tissue, as a leaf incubated in wall-digesting enzymes is left with pustules attached to vascular bundles when all mesophyll has been lost.

The localized unloading can be considered in terms of gradients of turgor pressure. The calculated pressure gradient for phloem transport in young barley is 0.5 MPa m^{-1}, with a volume flux of 0.45 m s^{-1}. Across an unloading distance of (say) 2 mm from vein to fungus, a turgor pressure difference of 0.35 MPa may exist (measured in the epidermis; Berryman *et al.*, 1991). With a hydraulic conductance across plasmodesmata of 2×10^{-4} m MPa s^{-1}, a volume flux out of phloem of 0.07 m s^{-1}, about 15% of that along phloem, would be generated, which is more than enough to provide sufficient sugars for fungal growth.

8.9 An overview of the carbon balance of infected plants

Implicit in this chapter has been the idea that the effect of biotrophic pathogens on the carbon balance of their hosts can be explained by the same principles that govern the carbon balance of healthy plants. No new concepts are needed; no special mechanisms operate. Changes in respiration and translocation are mediated not by the production of discrete and specific modulators, or toxins by the fungus (or by plant growth regulators), but simply by factors affecting the flux through existing machinery. It is not at all clear whether changes in amount of host protein, of SPS, acid invertase, membrane ATPases, and respiratory enzymes, are a consequence of altered flux, or of an as yet undefined fungus-specific cause. The simpler mechanism, which should be investigated first, is the former. By contrast, the factors that determine susceptibility and resistance in general seem far more specific and precise. This is not surprising: overwhelmingly, one plant genotype is similar to the next as a nutritional source for a potential pathogen, whilst the host range of most pathogens is exceedingly narrow. It remains to be seen if adult plant or race non-specific resistance has a nutritional component, unlike the better-studied major gene resistance. There is every reason to believe that a variety of pests and pathogens will interact in a fundamentally similar manner with the physiology of their hosts: our understanding of host physiology probably is the limiting factor.

Acknowledgements

I would like to thank P.E.H. Minchin and M.R. Thorpe for their hospitality in DSIR, Gracefield, New Zealand, during the writing of this chapter, and the Wain Fund of the AFRC for financing travel there.

References

Ahmad, I., Owera, S.A.P., Farrar, J.F. and Whitbread, R. (1982) The distribution of five major nutrients in barley plants infected with brown rust. *Physiol. Plant Pathol.* **21,** 335–346.

Ahmad, I., Farrar, J.F. and Whitbread, R. (1984) Fluxes of phosphorus in leaves of barley infected with brown rust. *New Phytol.* **99,** 361–375.

Al-Khesraji, T.O., Losel, D.M. and Gay, J.L. (1980) The infection of vascular tissue in leaves of *Tussilago farfara* L. by pycnial-aecial stages of *Puccinia poarum* Niels. *Physiol. Plant Pathol.* **17,** 193–197.

Baker, D.A. and Milburn, J.A. (1989) *Transport of Photoassimilates.* Longman, Harlow.

Battistelli, A., Adcock, M.D. and Leegood, R.C. (1991) The relationship between the activation state of sucrose-phosphate synthase and the rate of CO_2 assimilation in spinach leaves. *Planta* **183,** 620–622.

Berryman, C.A., Eamus D. and Farrar, J.F. (1991a) Water relations of leaves of barley infected with brown rust. *Physiol. Mol. Plant Pathol.* **38,** 393–405.

Berryman, C.A., Eamus D. and Farrar, J.F. (1991b) Variations in epidermal cell turgor of rust-infected barley seedlings. *New Phytol.* **119,** 535–540.

Berryman, C.A., Eamus, D. and Farrar, J.F. (1991c) The hydraulic conductivity of roots of rust-infected barley seedlings. *Physiol. Mol. Plant Pathol.* **38,** 407–415.

Bonnemain, J.L., Delrot, S., Lucas, W.J. and Dainty, J. (1991) *Recent Advances in Phloem Transport and Assimilate Compartmentation.* Ouest Editions, Nantes.

Borland, A.M. and Farrar, J.F. (1989) The partitioning of photosynthetically fixed carbon in the leaf blade and leaf sheath of *Poa pratensis* L. *J. Exp. Bot.* **40,** 1247–1254.

Bushnell, W.R. (1970) Patterns in the growth, oxygen uptake, and nitrogen content of single colonies of wheat stem rust on wheat leaves. *Phytopathol.* **60,** 92–99.

Bushnell, W.R. and Allen, P.J. (1962) Respiratory changes in barley leaves produced by single colonies of powdery mildew. *Plant Physiol.* **37,** 751–758.

Cairns, A.J. and Pollock, C.J. (1988) Fructan biosynthesis in excised leaves of *Lolium temulentum* L. II. Changes in fructosyl transferase activity following excision and application of inhibitors of gene expression. *New Phytol.* **109,** 407–413.

Chakravorty, A.K. and Scott, K.J. (1982) Biochemistry of host–rust interactions. In: *The Rust Fungi* (eds K.J. Scott and A.K. Chakravorty). Academic Press, London, pp. 180–223.

Daly, J.M. (1976) The carbon balance of diseased plants: changes in respiration, photosynthesis and translocation. In: *Encyclopedia of Plant Physiology*, Volume 4, *Physiological Plant Pathology* (eds R. Heitefuss and P.A. Williams). Springer, Berlin, pp. 450–479.

Day, D.A., De Vos, O.C., Wilson, D. and Lambers, H. (1985) Regulation of respiration in leaves and roots of two *Lolium perenne* populations with contrasting mature leaf respiration rates and crop yields. *Plant Physiol.* **78,** 678–683.

Delrot, S. (1989) Loading of photoassimilates. In: *Transport of Photoassimilates* (eds D.A. Baker and J.A. Milburn). Longman, Harlow.

Douce, R. (1985) *Mitochondria in Higher Plants.* Academic Press, New York.

Farrar, J.F. (1991) Starch turnover: its role in source–sink relations and a comparison with the role of sucrose. In: *Recent Advances in Phloem Transport and Assimilate Compartmentation* (eds J.I. Bonnemain, W. Lucas, S. Delrot and J. Dainty). Ouest Editions, Nantes, pp. 213–223.

Farrar, S.C. and Farrar, J.F. (1985) Carbon fluxes in leaf blades of barley. *New Phytol.* **100,** 271–283.

Farrar, S.C. and Farrar, J.F. (1986) Compartmentation and fluxes of sucrose in intact leaf blades of barley. *New Phytol.* **103,** 645–657.

Farrar, J.E. and Lewis, D.H. (1987) Nutrient relations in pathogenic and mutualistic infections. In: *Fungal Infection of Plants* (eds P.G. Ayres and G.F. Pegg). Cambridge University Press, pp. 92–132.

Farrar, J.F. and Rayns, F.W. (1987) Respiration of leaves of barley infected with powdery mildew: increased engagement of the alternative oxidase. *New Phytol.* **107,** 119–125.

Farrar, J.F. and Williams, J.H.H. (1990) Control of the rate of respiration in roots: compartmentation, demand, and the supply of substrate. In: *Compartmentation of Plant Metabolism in Non-Photosynthetic Tissues*, S. E. B. Seminar Series (ed. M. Emes). Cambridge University Press, pp. 167–188.

Farrar, J.F., Van der Schoot, C., Drent, P. and van Bel, A. (1992) Symplastic transport of lucifer yellow in mature leaf blades of barley: potential mesophyll-to-sieve tube transfer. *New Phytol.* **120,** 191–196.

Franceschi, V.R. and Giaquinta, R.T. (1983) The paraveinal mesophyll of soybean leaves in relation to assimilate transfer and compartmentation. *Planta* **157,** 422–431.

Haigh, G.R., Carver, T.L.W., Gay, A.P. and Farrar, J.F. (1991) Respiration and photosynthesis in oats exhibiting different levels of partial resistance to *Erysiphe graminis* f. sp. *avenae* Marchal. *New Phytol.* **119,** 129–136.

Huber, S.C. (1986) Fructose 2,6-bisphosphate as a regulatory metabolite in plants. *Ann. Rev. Plant Physiol.* **37,** 233–246.

Kneale, J. and Farrar, J.F. (1985) The localisation and frequency of haustoria in colonies of brown rust on barley leaves. *New Phytol.* **101,** 495–505.

Klecan, A.L. and Buchanan, B.B. (1988) Powdery mildew infection alters fructose-2-6-bisphosphate content and sucrose to starch ratio in leaves of wheat plants actively forming grain. *Physiol. Mol. Plant Pathol.* **32,** 221–227.

Klecan, A.L., Elliot, V.J. and Buchanan, B.B. (1986) A recognition metabolite for biological stress in plants: fructose-2,6-bisphosphate. *C.R. Acad. Sci. Paris* **302,** 51–54.

Komor, E. (1982) Transport of sugar. In: *Encyclopedia of Plant Physiology* vol. 13A, *Plant Carbohydrates I* (eds F.A. Loewus and W. Tanner). Springer, Berlin, pp. 635–676.

Krömer, S. and Heldt, K.W. (1991) On the role of mitochondrial oxidative phosphorylation in photosynthetic metabolism as studied by the effect of oligomycin on photosynthesis in protoplasts and leaves of barley. *Plant Physiol.* **95,** 1270–1276.

Lambers, H. (1985) Respiration in intact plants and tissues. In: *Encyclopedia of Plant Physiology,* vol. 18, *Respiration of Higher Plant Cells* (eds R. Douce and D.A. Day). Springer, Berlin, pp. 418–473.

Last, F.T. (1962) Analysis of the effects of *Erysiphe graminis* DC on the growth of barley. *Ann. Bot.* **26,** 279–289.

Madore, M.A. and Lucas, W.J. (1987) Control of photoassimilate movement in source-leaf tissues of *Ipomoea tricolor* Cav. *Planta* **171,** 197–204.

Manners, J.G. and Gay, J.L. (1983) The host–parasite interface and nutrient transfer in biotrophic parasitism. In: *Biochemical Plant Pathology* (ed. J.A. Callow). Wiley, Chichester, pp. 163–195.

Maynard, J.W. and Lucas, W.J. (1982) A reanalysis of the two component phloem loading system in *Beta vulgaris. Plant Physiol.* **69,** 734–739.

McAinsh, M.R., Ayres, P.G. and Hetherington, A.M. (1989) The respiration of protoplasts from leaves of barley infected by powdery mildew (*Erysiphe graminis* f. sp. *hordei*). *Plant Sci.* **64,** 221–230.

McAinsh, M.R., Ayres, P.G. and Hetherington, A.M. (1991) The effects of infection by powdery mildew (*Erysiphe graminis* f. sp. *hordei*) and low temperature on the respiratory activity of winter barley. *Physiol. Mol. Plant Pathol.* **39,** 13–23.

Murray, A.J.S. and Ayres, P.G. (1986) Uptake and translocation of nitrogen by mildewed barley seedlings. *New Phytol.* **104,** 355–365.

Owera, S.A.P., Farrar, J.E. and Whitbread, R. (1983) Translocation from leaves of barley infected with brown rust. *New Phytol.* **94,** 111–123.

Patrick, J.W. (1991) Control of phloem transport to and short-distance transfer in sink regions: an overview. In: *Recent Advances in Phloem Transport and Assimilate Compartmentation* (eds J.L. Bonnemain, S. Delrot, W.J. Lucas and J. Dainty). Ouest, Nantes, pp. 167–177.

Pilbeam, C.J., Robson, M.T. and Lambers, H. (1986) Respiration in mature leaves of *Lolium perenne* as affected by nutrient supply, cutting and competition. *Physiol. Plant.* **66,** 53–57.

Pollock, C.J. (1986) Fructans and the metabolism of sucrose in vascular plants. *New Phytol.* **104,** 1–24.

ap Rees, T. (1974) Pathways of carbohydrate breakdown in higher plants. In: *International Review of Biochemistry XI* (ed. D.H. Northcote). Butterworth, London, pp. 89–127.

Reinhold, L. and Kaplan, A. (1984) Membrane transport of sugars and amino acids. *Ann. Rev. Plant Physiol.* **35,** 45–83.

Roberts, A.M. and Walters, D.R. (1988) Photosynthesis in discrete regions of leeks infected with the rust *Puccinia allii* Rudd. *New Phytol.* **110,** 371–376.

Scheibe, R. (1991) Redox-modulation of chloroplast enzymes. *Plant Physiol.* **96,** 1–3.

Scholes, J.D. and Farrar, J.F. (1985) Photosynthesis and chloroplast functioning within individual pustules of *Uromyces muscari* on bluebell leaves. *Physiol. Plant Pathol.* **27,** 387–400.

Scholes, J.D. and Farrar, J.F. (1986) Increased rates of photosynthesis in localised regions of a barley leaf infected with brown rust. *New Phytol.* **104,** 601–612.

Smith, S.E. and Smith, F.A. (1990) Structure and function of the interfaces in biotrophic symbioses as they relate to nutrient transport. *New Phytol.* **114,** 1–38.

Stitt, M. (1987) Fructose 2,6-bisphosphate and plant carbohydrate metabolism. *Plant Physiol.* **84,** 201.

Streuter, N., Moerschbacher, B. and Reisener, H.J. (1989) Fructose 2,6-bisphosphate in wheat infected with stem rust. *J. Plant Physiol.* **134,** 165–168.

Sung, S.S., Xu D.-P., Galloway, C.M. and Black, C.C. (1988) A reassessment of glycolysis and gluconeogenesis in higher plants. *Physiol. Plant.* **72,** 650–654.

Tetlow, I.J. and Farrar, J.F. (1992) Sucrose-metabolising enzymes from leaves of barley infected with brown rust. *New Phytol.* **120,** 475–480.

van Bel, A.J.E. (1987) The apoplast concept of phloem loading has no universal validity. *Plant Physiol. Biochem.* **25,** 677–686.

van't Hof, J. and Kovacs, C.J. (1972) Mitotic cycle regulation in the meristem of cultured roots: the principal control point hypothesis. In: *The Dynamics of Meristem Cell Populations* (ed. M.W. Miller and C.C. Kuehnert). Plenum, New York, pp. 15–32.

Walters, D.R. (1985) Shoot : root inter-relationships: the effects of obligate biotrophic fungal pathogens. *Biol. Rev.* **60,** 47–79.

Walters, D.R. and Ayres, P.G. (1981) Growth and branching pattern of roots of barley infected with powdery mildew. *Ann. Bot.* **47,** 159–162.

Walters, D.R. and Ayres, P.G. (1982) Translocation of ^{14}C-labelled photoassimilates to roots in barley: effects of mildew on partitioning in roots and the mitotic index. *Plant Pathol.* **31,** 307–313.

Wardlaw, I.F. (1990) The control of carbon partitioning in plants. *New Phytol.* **116,** 341–381.

Whipps, J.M. and Lewis, D.H. (1981) Patterns of translocation, storage and interconversion of carbohydrates. In: *Effects of Disease on the Physiology of the Growing Plant* (ed. P.G. Ayres). Cambridge University Press, pp. 47–83.

Williams, J.H.H., Minchin, P.E.H. and Farrar, J.F. (1991) Carbon partitioning in split root systems of barley: the effect of osmotica. *J. Exp. Bot.* **432,** 453–460.

Williams, M.L., Farrar, J.F. and Pollock, C.J. (1989) Cell specialisation in leaves of *Hordeum distichum* L.: the parenchymatous bundle sheath. *Plant, Cell Environ.* **12,** 909–918.

Wilson, J.B. (1989) A review of evidence on the control of shoot : root ratio, in relation to models. *Ann. Bot.* **61,** 433–449.

Zulu, J.N., Farrar, J.F. and Whitbread, R. (1991a) Effects of phosphate supply on the phosphorus status, dry mass and photosynthesis of wheat infected with powdery mildew. *New Phytol.* **118,** 453–462.

Zulu, J.N., Farrar, J.F. and Whitbread, R. (1991b) Effects of phosphate supply on wheat infected with powdery mildew: carbohydrate metabolism of source leaves. *New Phytol.* **118,** 553–558.

9

Partitioning to storage, regrowth and reproduction

N.D. Paul

9.1 Introduction

Infestation by foliar pests and infection by foliar pathogens both result in a more or less localized dysfunction of shoot tissues. In seeking to compare whole-plant responses to pests and pathogens it is tempting to seek analogies between broad categories of pathogen and herbivore. For example, are the effects of tissue death caused by necrotrophic pathogens broadly similar to those resulting from the removal of tissue by biting or chewing herbivores? Equally, what parallels are there between the localized dysfunction of living tissues infected by biotrophic pathogens and those attacked by phloem-feeding invertebrates? As yet, however, our understanding of the physiological consequences of particular forms of disease or herbivory are insufficient for such questions to be addressed and we are limited to considering responses to pathogens on the one hand and to herbivores on the other. There is considerable evidence which demonstrates that many whole-plant responses are common to perturbations in shoot function induced by a range of abiotic factors as well as by herbivores or disease (Ayres, 1984a). On this basis we might expect close parallels between whole-plant responses to herbivory and disease. This chapter assesses whether the literature reveals these anticipated parallels.

At the outset it is worth noting two caveats. Firstly, even the most cursory examination of the literature reveals great differences between the perceptions of those working with herbivores and those working with pathogens, in terms of the processes and responses which they study. Thus, comparisons made here of the effects of herbivores and pathogens on a particular process are usually based on a very small number of investigations. Secondly, although storage, regrowth and reproduction are convenient subheadings under which to consider the effects of foliar pests or pathogens, such a division is unlikely to apply in the plant, so their integration must also be considered. This is done here, for diseased plants, by reference to a model system, *Senecio vulgaris* (groundsel) infected by the rust fungus, *Puccinia lagenophorae*

9.2 Storage and regrowth

Most studies of the physiological consequences of foliar pathogens have utilized relatively short-lived, annual crops in experiments lasting a few weeks at most. Thus, while the effects of disease or herbivory on storage and regrowth may be of greatest significance in biennial or perennial hosts, assessment of such effects in longer-lived hosts must unfortunately rely on extrapolation from short-term studies of annuals.

As has been highlighted by Scholes (Chapter 7), a range of pathogens, both necrotrophic and biotrophic, reduce the rate of carbon fixation in infected tissues. Even if compensatory increases in photosynthesis occur in uninfected tissues (e.g. Murray and Walters, 1992), reduced fixation in infected tissues, coupled with increased respiratory losses and sequestration of photosynthate by the pathogen (see Farrar, Chapter 8) tend to reduce the supply of carbohydrate available to host sinks, including storage tissues. In barley, *Hordeum vulgare*, reduced photosynthesis and assimilate export caused by infection with powdery mildew, *Erysiphe graminis*, results in the depletion of root carbohydrates, especially the soluble fraction, within days of infection (Walters, 1981). The concentration of soluble carbohydrates also declined in the roots of *Poa pratensis* affected by flag smut, *Urocystis agropyri* (Hodges and Robinson, 1977). By contrast, infection of the same host by stripe smut, *Ustilago striiformis*, caused no reduction in soluble carbohydrate in roots (Hodges and Robinson, 1977). The contrast between these two smuts was also apparent in the far greater reduction in dry weight partitioning to rhizomes caused by *U. agropyri* (Nus and Hodges, 1990).

The reduced partitioning to rhizomes in smutted *P. pratensis* reflects a reduced translocation to below-ground organs, which is usually evident in a reduction of the dry weight root:shoot ratio (R:S) caused by foliar infections (Walters, 1984). Similar decreases in R:S also occur in response to other factors which inhibit shoot function (Ayres, 1991), including herbivory (see Chapter 2). Obviously, decreased R:S resulting from reduced export from the shoots of grazed plants may be confounded in the short term by the direct removal of shoot tissue by herbivores. Similarly, transient increases in R:S may also occur in plants infected by necrotrophic pathogens which cause the loss of shoot tissue.

The consequences of reduced R:S for the diseased plant have generally been considered in terms of alterations in the functional balance between root and shoot (Walters, 1984; Ayres, 1984b). The other roles of the root system, for example as a source of storage reserves, have generally been ignored. However, studies of shoot herbivory highlight the need to consider the possible long-term effects of reduced root growth.

Reduced root development in the perennial herb, *Rumex crispus*, as a result of grazing by a chrysomelid beetle, *Gastrophysa viridula*, resulted in increased vulnerability to flooding (Whittaker, 1982) and competition (Bentley and Whittaker, 1979). By contrast, *Rumex obtusifolius* showed little change in root weight following *Gastrophysa* grazing and this difference in tolerance is reflected in the response of the host to subsequent environmental stress (Bentley and Whittaker, 1979). Both *Rumex* species are infected by the biotrophic rust fungus *Uromyces rumicis* and the necrotrophs *Ramularia rubella* and *Venturia rumicis* (Ellis and Ellis, 1985). Although investigations of these pathogens are few there is evidence that their effects may parallel those of *G. viridula*. Thus, root growth in *R. obtusifolius* was reduced by approximately 60% by *R. rubella* but R:S was little changed (Huber-Meinicke *et al.*, 1989) while, in *R. crispus*, infection by *U. rumicis* reduced regrowth after defoliation (Frank, 1973). It is predictable that these fungi will have differential effects on *R. crispus* and *R. obtusifolius* comparable to those of *G. viridula* grazing, but this possibility remains uninvestigated.

The stability of root growth in *R. obtusifolius* under herbivore and, perhaps, pathogen attack may be a relatively widespread feature of wild plants. Ben-Kalio and Clarke (1979) reported that the ruderal annual *Senecio vulgaris* (groundsel) maintained partitioning to the roots despite severe infection by powdery mildew, *Erysiphe fischeri*. They argued that the maintenance of dry-weight partitioning patterns represented a mechanism of tolerance to mildew which was poorly expressed in most crop plants. There is also substantial variation in the magnitude of changes in R:S to herbivory, as in *Rumex* spp. attacked by *G. viridula* (Bentley and Whittaker, 1979). It has been argued elsewhere (Paul, 1989, 1990) that such between-species variability in the responses of partitioning to stress are part of the broader adaptation of the plant to its environment, and that changes in R:S in response to one factor, such as shading, will predict those to others, including disease. This hypothesis predicts that in a given host species different consumers will tend to have similar effects on R:S, but this need not always be the case. For example, partitioning in *S. vulgaris* shows less stability following infection by rust than powdery mildew (see below), while Clarke (1984) suggests that cereals may be more tolerant of rusts than powdery mildews. It is possible that this apparent variation may depend partly upon different host genotypes being used in different studies, but this cannot account for contrasting effects of different smuts on partitioning to rhizomes in *P. pratensis* (see above). Clearly, patterns of change in partitioning are determined by more subtle alterations in host physiology than can be summarized as 'shoot dysfunction'.

9.3 Reproduction

9.3.1 *Changes in dry-weight partitioning to reproduction*

Foliar pathogens of cereals and other seed crops are generally considered to reduce dry-weight partitioning to yield, i.e. to reproduction. Pathogen infection also reduces partitioning to reproduction in some wild plants (e.g. Paul and Ayres, 1987; DeNooij, 1987). However, some caution is required when interpreting changes in partitioning to reproduction. Thus, while partitioning to reproduction in groundsel is lower throughout development in plants infected by rust (Figure 9.1a), the allometric relationship between reproduction and plant weight is not significantly affected by rust (Figure 9.1b). Expressed in this manner, it can be argued that partitioning to reproduction in groundsel is well buffered against disease, the observed changes being the result of reduced plant size. It is worth noting that this stability in the allometry

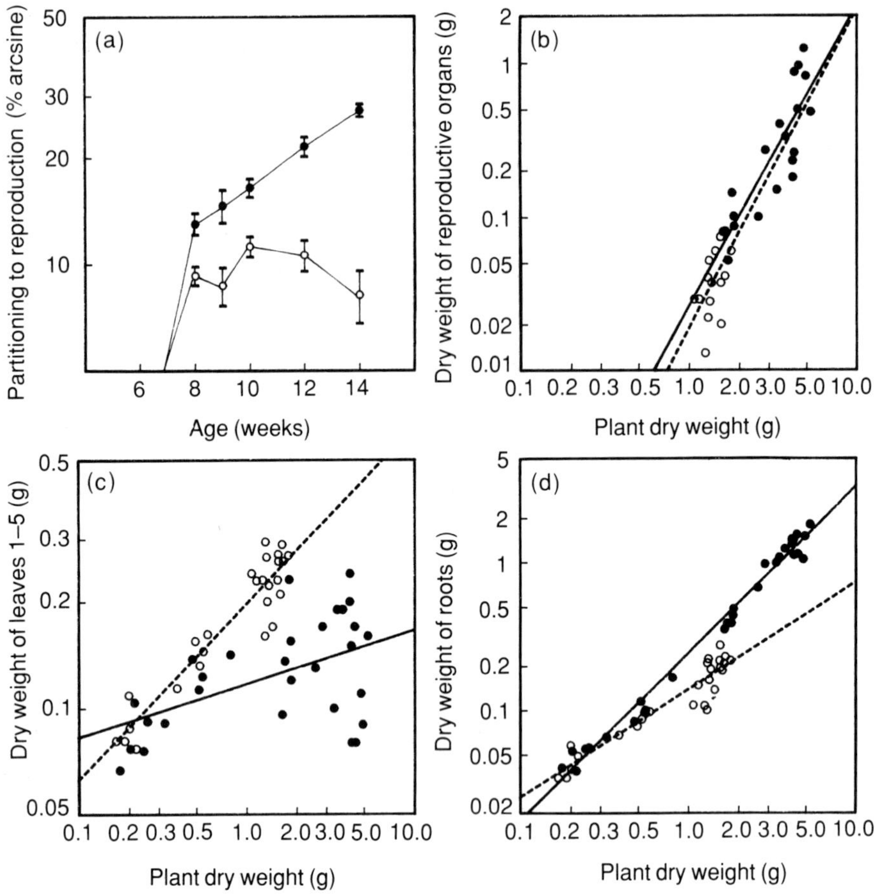

***Figure 9.1**. Dry-weight partitioning in* Senecio vulgaris *uninfected and infected by the rust fungus* Puccinia lagenophorae *when 4 weeks old. Control plants (●); infected plants (○). (a) The partitioning of dry weight to reproduction during development. Note: data have been arcsine transformed. Bars are two s.e. of the mean of five replicates. (b) The allometric relationship between reproductive dry weight and total plant dry weight. Allometric constants were: control 2.27 ± 0.13, rusted 1.97 ± 0.19 (not significantly different). (c) The allometric relationship between the dry weight of leaves 1–5 and total plant dry weight (leaf 1 is the first true leaf and plants were infected when these leaves were approaching full expansion). Allometric constants were: control 0.15 ± 0.05, rusted 0.40 ± 0.08 ($P < 0.001$). (d) The allometric relationship between root dry weight and total plant dry weight. Allometric constants were: control 1.13 ± 0.03, rusted 0.66 ± 0.16 ($P < 0.01$).*

of partitioning to reproduction contrasts with the shift in partitioning between leaves and roots which, in groundsel at least, is evident in allometric relationships as well as in measurements of R:S (Figure 9.1 c,d)

9.3.2 *Changes in reproductive output*

Clearly, changes in dry-weight partitioning to reproduction are remote from seed output and, ultimately, fitness, where this is the number of offspring produced. Foliar disease can reduce the number of seeds produced (Inman, 1971; Burdon, 1982; Alexander and Burdon, 1984; Paul and Ayres, 1986a; DeNooij, 1987) and infection may also delay the start of flowering (Paul and Ayres, 1986a). Plants in which flowering is delayed may be less efficiently pollinated and, since seeds which germinate early may gain a persistent advantage over those developing later (e.g. Millar, 1987), such a pathogen-induced delay in flowering may have effects which persist into the next generation. Parental infection may also affect the next generation if seed quality is reduced. Disease may reduce seed weight, as in *R. crispus* infected by *U. rumicis* (Inman, 1971), and lighter seed tend to produce relatively slow-growing seedlings which perform poorly in competition with seedlings from undamaged parents (Cidiceyan and Malloch, 1982). Parental disease may also have longer term effects on the fitness of offspring, as with *Linum marginale* in which seedlings of parents infected by *Melampsora lini* showed substantial reductions in fecundity (Burdon and Jarosz, 1988). Again, such responses may follow herbivory as well as disease, for example grazing by *G. viridula*, like rust infection, reduces seed weight in *R. crispus* (Bentley *et al.*, 1980).

Even when seed production and/or quality is reduced by disease this need not necessarily lead to reduced host population density. *Spartina anglica* is highly susceptible to floral infection by *Claviceps purpurea* which reduces fecundity 'by 16% on average . . . and by nearly half in some populations' (Gray *et al.*, 1989). However, this loss has little effect on *Spartina* populations since seed production is of minor significance relative to vegetative spread (Gray *et al.*, 1989). A similar phenomenon is apparent in a more extreme form in plants infected by systemic endophytes which sterilize their hosts but enhance fitness through increased vegetative vigour and clonal reproduction (Clay, 1990). A more general case in which seed output may be reduced without modifying host population density is where seedling recruitment is limited by the availability of suitable sites for germination rather than by the number of seed (e.g. Andersen, 1989). Such limits are frequently cited where apparently 'effective' biocontrol agents have failed to limit natural weed populations in the field. For example, infestation by seed head flies, *Urophora* spp., reduces seed production in *Centaurea diffusa* and *C. maculosa* by approximately 90% but 'the quantity of seed not destroyed is more than adequate to maintain knapweed populations. . . .' (Cloutier and Watson, 1989).

Prins and Verkaar (see Chapter 2) note that the absence of studies into the effects of herbivory on male fitness limits our understanding of the effects of grazing on host reproduction. It is equally true that the effects of disease on male as well as female reproductive fitness should be considered. In dioecious plants, males are subject to more severe attack by both pathogens and herbivores but suffer lower mortality as a result (Agren, 1987). However, little is known of the effects of foliar disease on male reproductive fitness. Defoliation may reduce nectar production, as in *Epilobium angustifolium* (Michaud, 1989), although no such response was found in *Polemonium foliosissimum* (Zimmerman and Pyke, 1988). It seems likely that the reduced phloem export from rust-infected leaves (Farrar, Chapter 8) might also significantly reduce nectar production but this remains uninvestigated. In *Ambrosia artemisiifolia* the number of male flowers and quantity of pollen shed was greatly reduced in plants systemically infected by *Albugo tragopogonis* (Hartmann and Watson, 1980). Interestingly, the same study indicated that plants with localized infections tended to produce more pollen than controls (46.9 $\pm$ 80.2 and 22.0 $\pm$ 21.0 million grains per plant, respectively) whereas seed production was little changed (75.8 $\pm$ 55.2 and 98.7 $\pm$ 71.8 seeds per plant in infected and control, respectively). The data suggest that we cannot assume that male and female reproductive fitness will respond equally to foliar damage.

9.3.3 *Can foliar infection result in increased reproduction?*

The question of 'overcompensation' following herbivory has stimulated much debate (see Chapter 2). While this subject remains controversial, there seems to be some agreement that grazing may increase reproductive success, at least in some species under certain conditions (Maschinsky and Whitham, 1989). The question of overcompensation does not seem to have been addressed by pathologists. However, some insight can be gained from those root-infecting fungi which form vesicular–arbuscular mycorrhizas. Uptake of phosphorus via these organisms has been found in many host species (Harley and Smith, 1983), but infection may substantially inhibit growth of the host during certain periods of its development (Koide, 1985) or under particular environmental conditions, for example high phosphorus and low light (Son and Smith, 1988). The physiological basis for this dichotomy is well known, the benefit of increased phosphate uptake being set against the cost of its acquisition in terms of carbon and energy expended by the fungus (Fitter, 1991). It is difficult to imagine a comparable balance between costs and benefits in plants arising from most leaf-infecting fungi. One notable exception is infection by foliar endophytes, which may benefit the host, for example by deterring herbivory (Clay, 1990). Endophytes may also inhibit or wholly suppress sexual reproduction but favour vegetative vigour, so increasing fitness in terms of spread and persistence in swards (Clay, 1988, 1990). It is worth noting that such effects may result from fungi which are normally considered as pathogenic, for example *Epichloe typhini* (choke) in *Agrostis tenuis* (Bradshaw, 1959). Even more unexpected is the increased vegetative vigour and survivorship of *Pulsatilla pratensis* infected by *Puccinia pulsatillae*, a systemic rust in the same genus as many damaging pathogens (Wennstrom and Ericson, 1991). This result, while involving a shift from seed to vegetative reproduction, rather than an increase in seed production, does suggest that the possibility of increased host fitness resulting from fungi normally considered pathogens should not be excluded.

9.4 Integration of storage, regrowth and reproduction: a model system

Groundsel is a ruderal annual which may germinate throughout the year when conditions permit. Plants germinating in spring and summer grow rapidly and reproduction commences within 8–10 weeks of germination. By contrast, seedlings developing in autumn form overwintering rosettes which may not flower until the following March or April. The rust fungus, *P. lagenophorae*, may infect plants throughout the year and causes a variety of physiological dysfunctions which result in reduced growth and reproduction in the field (Paul and Ayres, 1986a, 1987).

9.4.1 *Changes in partitioning and storage*

In rusted plants, dry weight is retained in infected leaves at the expense of roots and stems (Paul and Ayres, 1986b). Root growth is reduced in rusted plants throughout the life of the host but this inhibition becomes particularly marked after the onset of flowering (Paul and Ayres, 1986b). In control plants, root dry-weight and length continue to increase for several weeks after flowering begins, whereas in rusted plants dry-weight accumulation in the roots virtually ceases at that time (Paul and Ayres,1986c). Confirmation that the cessation of dry weight increase in the roots of rusted plants represents a failure to accumulate storage compounds (Paul and Ayres, 1986c) has come from recent examination of the effects of rust infection on the carbohydrate relations of *S. vulgaris*. Non-structural carbohydrates (NSC) were measured following the method of Farrar (1980) which defines three pools, soluble NSC (extractable in hot 95% ethanol), starch and a fructan fraction (extractable in cold acetate buffer).

Starch and, especially, fructans accumulated in infected leaves while soluble sugars declined. Comparable accumulation of starch and fructans have been reported in a range of hosts infected by foliar biotrophs (Whipps and Lewis, 1981; Farrar, 1989; Zulu *et al.*, 1991). It is not possible to quantify separately the storage carbohydrates in host and fungus, although fungal glycogen as well as host starch accumulates in *Tussilago farfara* leaves infected by *Puccinia poarum* (Holligan *et al.*, 1973). However, there is no question that much of the carbohydrate accumulated in infected leaves is within host tissues (see Scholes, Chapter 7; also Farrar, Chapter 8) and so, in principle, may be partly available for remobilization if sink demand elsewhere in the plant increases. The possible magnitude of such remobilization, and whether it occurs at all, remains uninvestigated in infected plants.

Accumulation of NSC in infected tissues (Figure 9.2a) is at the expense of uninfected roots and stems (Figure 9.2b,c). In the roots of groundsel, the concentration of NSC, especially soluble sugars and fructans, increased progressively prior to the onset of reproduction (10 weeks of age), at which time the content of NSC in the roots of rusted plants was approximately half that in controls. NSC storage was similarly reduced in the stems (Figure 9.2c). In rusted plants, the reserves of NSC in the roots and stems were rapidly lost following the onset of flowering (Figure 9.2b,c). By contrast, control plants continued to accumulate NSC in stems until at least 13 weeks

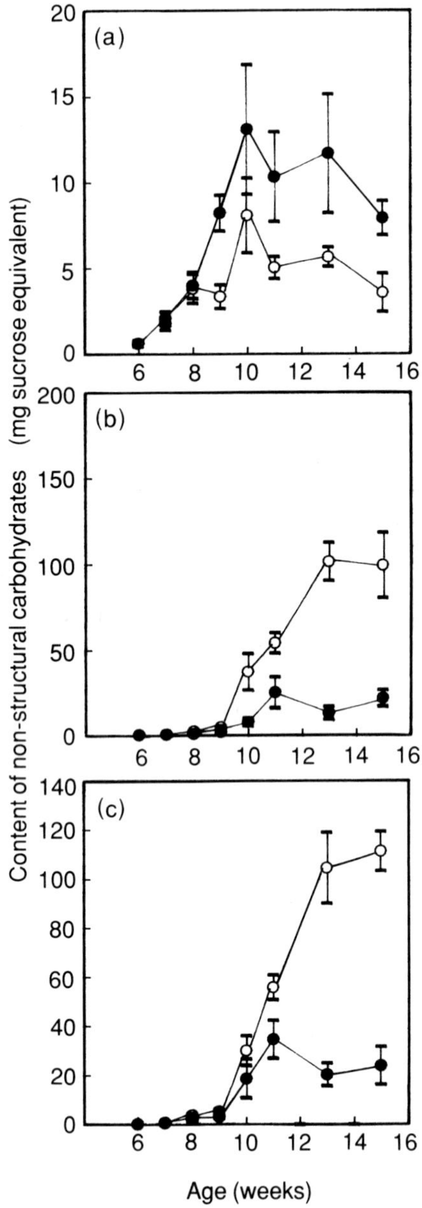

***Figure 9.2**. Contents of non-structural carbohydrates in (a) infected leaves, (b) roots and (c) stems of* Senecio vulgaris. *Control plants (○); plants infected with* Puccinia lagenophorae *when 4 weeks old (●). Bars are two s.e. of the mean of five replicates.*

of age (Figure 9.2c). There is some depletion of NSC reserves in the roots of controls, but this is smaller and occurs 2–4 weeks later than in rusted plants. Thus, it appears that rust-infected groundsel supports reproductive development largely on the basis of stored rather than current photosynthate. This is certainly consistent with the reduced period of flowering and reduced seed production caused by rust infection (Paul and Ayres, 1986a, 1987).

9.4.2 *Consequences of changed storage*

The depletion of storage compounds to support flowering presumably also compromises the ability of rusted groundsel to mobilize reserves in response to other environmental factors, such as defoliation. Even prior to flowering, rusted plants have a greater proportion of total reserves present in the shoot than the root. They may be vulnerable to any factor, biotic or abiotic, which causes a loss of shoot tissue, even assuming host carbohydrates accumulated in the vicinity of rust infections are available for remobilization by the host. This increased vulnerability of rusted plants to secondary stress is clearly demonstrated in overwintering groundsel. In young, cold-acclimated plants (12 h day; day temperature 8°C, night temperature 4°C) more than 60% of total NSC is present in the leaves of rusted plants, as opposed to less than 50% in healthy plants. In the field, this altered balance between root and shoot storage occurs in rusted plants at the onset of winter cold. Rusted leaves are particularly vulnerable to frost damage (Paul and Ayres, 1991) and infected plants suffer more leaf death than controls during winter (Paul and Ayres, 1986d). Clearly, increased loss of leaves, which hold the major part of the plant's NSC, results in very significant depletion of total reserves. Under controlled freezing treatments in the laboratory, leaf death in control plants was low and was more than compensated for by continuing leaf expansion. As a result, total NSC increased 2.4-fold during 4 weeks when night temperature was reduced to −5°C. By contrast, over the same period, leaf death in rusted plants resulted in the loss of over 50% of accumulated reserves.

Depletion of reserves has major effects on the regrowth of groundsel when environmental conditions improve in spring. The vulnerability of rusted plants to such depletion is likely to be increased by the reduced survival over winter of leaf area, which places increased importance on stored rather than current photosynthesis during initial recovery. Regrowth in rusted plants lags behind that of controls both under controlled conditions and in the field (Paul and Ayres, 1986c). Even though the destruction of infected tissues over winter renders the plants free of rust in spring, reproduction of previously rusted plants is delayed compared with controls and seed output is reduced (Paul and Ayres, 1986b). Thus, the effects of rust infection on storage during autumn have effects on regrowth and reproduction which persist long after infection has ceased.

Such losses are not peculiar to groundsel or other over wintering annuals, such as *Capsella bursa-pastoris* (Alexander and Burdon, 1984). Burdon and Jarosz (1988) reported that over winter mortality of perennial flax (*L. marginale*) was proportional to the severity of rust infection during the previous summer, and they cited comparable effects in *Phlox* spp. infected by *Erysiphe cichoracearum*. In *R. crispus*, inoculation with *U. rumicis* during summer reduced over winter survival from 95% to 43% in one study (Frank, 1973) and from 94% to approximately 40% in another (Inman, 1971). Such responses to infection may be representative of a range of biennial and perennial hosts in which over winter storage of reserves in perennating organs is a key process.

Outbreaks of foliar pathogens are often most severe during late summer and autumn, when cooler, moister conditions favour spore dispersal and infection. Such late infections may have little effect on the current season's growth or reproduction but, if they coincide with periods of accumulation in storage organs, the host may be compromised in the following season, long after any direct evidence of infection has disappeared.

Frost damage is not the only factor which may result in the preferential loss of NSC-rich rusted leaves. A range of invertebrate herbivores has been found to graze preferentially on rusted tissues (e.g. Ramsell and Paul, 1990; and see Chapter 10). As a result of the accumulation of storage carbohydrates and other resources in infected tissues, losses due to herbivory of infected tissues may be greater than those from an equivalent area of healthy tissue. For example, in *Tussilago farfara* grazed by the slug *Arion hortensis*, infection by *Puccinia poarum* increased the removal of leaf area by 3.7-fold: because of nitrogen accumulation in rusted tissues, this change was equivalent to a 9-fold increase in the loss of nitrogen (Ramsell and Paul, 1990). Similarly, rust infection would have increased loss of carbohydrate by a factor of nine in the light, and as much as 24-fold in the dark (Holligan *et al.* 1973), and loss of lipids by a factor of 13 (Losel and Lewis, 1974). Clearly, the cost of this loss of reserves must be set against the potential benefit of the removal of the pathogen, but the scope for a diversity of interactions between pathogens and non-vector herbivores is clearly substantial (see Chapter 10).

9.5 Conclusion

As noted in the Introduction, the focus of investigations into storage, regrowth and reproduction in grazed plants is rather different from that in studies of diseased plants. Those working with herbivores have generally considered integrated changes in growth, storage and reproduction in terms of the overall consequences of grazing on the ecology of the host: experimental approaches have often been relatively long term, utilizing a wide range of hosts and considering the consequences of grazing for host interactions with other factors. By contrast, those working with pathogens have usually sought a detailed understanding of changes in specific physiological processes in diseased plants: studies have tended to be rather short term and to use annual plants. The contrast between these different approaches is clear, but can the disjunct knowledge of disease and herbivory be integrated, allowing a general understanding of whole-plant responses to these consumers to be obtained?

As yet, the answer to that question must be no. Those studies which can be compared do reveal broad parallels between whole-plant responses to herbivory and disease. Such parallels might be anticipated but there are also indications of rather unexpected differences in response, as in *P. pratensis* infected by smut fungi (Nus and Hodges, 1990). That such closely related organisms can induce contrasting responses must indicate a need for caution in assuming comparability between foliar pests and pathogens. Clearly, only direct comparison of a range of consumers on the same host will clarify the limits of possible generalizations. The scope for such studies is enormous; many plants are host to a wide number of herbivores and pathogens.

Perhaps the greatest practical constraint will be to establish a common basis of quantifying the intensity of attack by such diverse organisms. Apart from their inherent interest, such studies will be essential in any attempt to integrate the interactions of plants with their pests and pathogens within generalized models of the possible consequences of climate change on plants and vegetation.

Acknowledgements

I am grateful to Paul Hatcher and Helen West for their comments on the manuscript and to NERC for their financial support of the groundsel rust studies.

References

Agren, J. (1987) Intersexual differences in phenology and damage by herbivores and pathogens in dioecious *Rubus chamaemorus* L. *Oecologia* **72**, 161–169.

Andersen A.N. (1989) How important is seed predation to recruitment in stable populations of long-lived perennials? *Oecologia* **81**, 310–315.

Alexander, H.M. and Burdon, J.J. (1984) The effect of disease induced by *Albugo candida* (white rust) and *Peronospora parasitica* (downy mildew) on the survival and reproduction of *Capsella bursa-pastoris*. *Oecologia* **64**, 314–318.

Ayres, P.G. (1984a) The interaction between environmental stress injury and biotic disease physiology. *Ann. Rev. Phytopathol.* **22**, 53–75.

Ayres, P.G. (1984b) Effects of infection on root growth and function; consequences for plant nutrient and water relations. In: *Plant Diseases, Infection, Damage and Loss* (eds R.K.S. Wood and G.J. Jellis). Blackwell Scientific Publications, Oxford, pp. 105–117.

Ayres, P.G. (1991) Growth responses induced by pathogens and other stresses. In: *Responses of Plants to Multiple Stresses* (eds H.A. Mooney, W.E. Winner and E.J. Pell). Academic Press, New York, pp. 227–248.

Bentley, S. and Whittaker, J.B. (1979) Effects of grazing by a chrysomelid beetle, *Gastrophysa viridula*, on competition between *Rumex obtusifolius* and *Rumex crispus*. *J. Ecol.* **67**, 79–90.

Bentley, S., Whittaker, J.B. and Malloch, A.J.C. (1980) Field experiments on the effects of grazing by a chrysomelid beetle (*Gastrophysa viridula*) on seed production and quality in *Rumex obtusifolius* and *Rumex crispus*. *J. Ecol.* **68**, 671–674.

Ben-Kalio, V.D. and Clarke, D.D. (1979) Studies on tolerance in wild plants: effects of *Erysiphe fischeri* on the growth and development of *Senecio vulgaris*. *Physiol. Plant Pathol.* **14**, 203–211.

Bradshaw, A.D. (1959) Population differentiation in *Agrostis tenuis* Sibth. II. The incidence and significance of infection by *Epichloe typhina*. *New Phytol.* **58**, 310–315.

Burdon, J.J. (1982) The effect of fungal pathogens on plant communities. In: *The Plant Community as a Working Mechanism* (ed. E.I Newman). Blackwell Scientific Publications, Oxford, pp. 99–112.

Burdon, J.J. and Jarosz, A.M. (1988) The ecological genetics of plant-pathogen interactions in natural communities. *Phil. Trans. R. Soc. B* **321**, 349–363.

Cidiceyan, M.A. and Malloch, A.J.C. (1982) Effects of seed size on the germination and competitive ability of *Rumex crispus* and *Rumex obtusifolius*. *J. Ecol.* **70**, 227–232.

Clarke, D.D. (1984) Tolerance of parasites and diseases in plants and its significance in host-parasite interactions. In: *Advances in Plant Pathology*, Volume 5 (eds D.S. Ingram and P.H. Williams). Academic Press, London, pp. 161–197.

Clay, K. (1988) Clavicipitaceous fungal endophytes of grasses: coevolution and the change from parasitism to mutualism. In: *Coevolution of Fungi with Plants and Animals* (eds D. Hawksworth and K. Pirozynski). Academic Press, London, pp. 79–105.

Clay, K. (1990) Clavicipitaceous endophytes of grasses: their potential as biocontrol agents. *Mycol. Res.* **92**, 1–12.

Cloutier, D. and Watson, A.K. (1989) Application of modelling to biological weed control. In: *Proceedings VII International Symposium on Biological Control of Weeds* (ed. E.S. Delfosse) CSIRO Publications, East Melbourne, pp. 6–11.

DeNooij, M.P. (1987) *On disease in natural populations of* Plantago lanceolata. Ph.D. Thesis, University of Utrecht.

Ellis, M.B. and Ellis, J.P (1985) *Microfungi on Land Plants.* Croom Helm, London.

Farrar, J.F. (1980) Allocation of carbon to growth, storage and respiration in the vegetative barley plant. *Plant Cell Environ.* **3**, 97–105.

Farrar, J.F. (1989) Fluxes and turnover of sucrose and fructans in healthy and diseased plants. *J. Plant Phys.* **134**, 137–140.

Fitter, A.H. (1991) Costs and benefits of mycorrhizas: implications for functioning under natural conditions. *Experientia* **47**, 350–355.

Frank, P.A. (1973) A biological control agent for *Rumex crispus* L. *Commonwealth Institute of Biological Control Miscellaneous Publication* **6**, 121–126.

Gray, A.J., Drury, M. and Raybould, A.F. (1989) *Spartina* and the ergot fungus *Claviceps purpurea* – a singular contest. In: *Pests, Pathogens and Plant Communities* (eds J.J. Burdon and S.R. Leather). Blackwell Scientific Publications, Oxford, pp. 63–79.

Harley, J.L. and Smith, S.E. (1983) *Mycorrhizal Symbiosis.* Academic Press, London.

Hartmann, H. and Watson, A.K. (1980) Damage to common ragweed (*Ambrosia artemesiifolia*) caused by the white rust fungus (*Albugo tragopogi*). *Weed Sci.* **28**, 632–635.

Hodges, C.F. and Robinson, P.W. (1977) Sugar and amino acid content of *Poa pratensis* infected by *Ustilago striiformis* and *Urocystis agropyri. Physiol. Plant.* **41**, 25–28.

Holligan, P.M., McGee, E.E.M. and Lewis, D.H. (1973) Quantitative determination of starch and glycogen and their metabolism in leaves of *Tussilago farfara* during infection by *Puccinia poarum. New Phytol.* **73**, 873–879.

Huber-Meinicke, G., Defago, G. and Sedlar, L. (1989) *Ramularia rubella* (Bon) Nannf as a potential mycoherbicide against *Rumex* weeds. *Bot. Hel.* **99**, 81–89.

Inman, R.E. (1971) A preliminary evaluation of *Rumex* rust as a biological control agent for curly dock. *Phytopathol.* **61**, 102–107.

Koide, R.T. (1985) The nature of growth reductions in sunflower caused by vesicular-arbuscular mycorrhizal infection. *New Phytol.* **99**, 449–462.

Losel, D.N. and Lewis, D.H. (1974) Lipid metabolism in leaves of *Tussilago farfara* during infection by *Puccinia poarum. New Phytol.* **73**, 1157–1169.

Maschinsky, J. and Whitham, T.G. (1989) The continuum of plant responses to herbivory: the influence of plant association, nutrient availability and timing. *Am. Nat.* **134**, 1–19.

Michaud, J.P (1989) Nectar accumulation in flowers of fireweed, *Epilobium angustifolium* (Onagraceae), in response to simulated defoliation. *J. Api. Res.* **28**, 181–186.

Millar, T.E. (1987) Effects of emergence time on survival and growth in an early old-field plant community. *Oecologia* **72**, 272–282.

Murray, D.C. and Walters, D.R. (1992) Increased photosynthesis and resistance to rust infection in upper, uninfected leaves of rusted broad bean (*Vicia faba* L.). *New Phytol.* **120**, 235–242.

Nus, J.L. and Hodges, C.F. (1990) Tiller and rhizome growth of water-stressed *Poa pratensis* 'Merion' infected by *Ustilago striiformis* or *Urocystis agropyri. Plant Dis.* **74**, 886–888.

Paul, N.D. (1989) Effects of fungal pathogens on the N, P and S relations of individual plants and populations. In: *Physiology and Ecology of Nitrogen, Phosphorus and Sulphur Utilisation by Fungi* (eds L. Boddy, R.Marchant and D.J. Read). Cambridge University Press, pp. 153–180.

Paul, N.D. (1990) Modification of the effect of plant pathogens by other components of natural ecosystems. In: *Pests, Pathogens and Plant Communities* (eds J.J. Burdon and S.R. Leather). Blackwell Scientific Publications, Oxford, pp. 81–96.

Paul, N.D. and Ayres, P.G. (1986a) Impact of a pathogen (*Puccinia lagenophorae*) on populations of groundsel (*Senecio vulgaris*) over wintering in the field. II. Reproduction. *J. Ecol.* **74**, 1085–1094.

Paul, N.D. and Ayres, P.G. (1986b) The effects of infection by rust (*Puccinia lagenophorae* Cooke) on the growth of groundsel (*Senecio vulgaris*) under a range of nutrient conditions. *Ann. Bot.* **58**, 321–331.

Paul, N.D. and Ayres, P.G. (1986c) The effects of nutrient deficiency and rust infection on the relationship between root dry weight and root length in groundsel (*Senecio vulgaris*). *Ann. Bot.* **57**, 353–360.

Paul, N.D. and Ayres, P.G. (1986d) Impact of a pathogen (*Puccinia lagenophorae*) on populations of groundsel (*Senecio vulgaris*) over wintering in the field. I. Mortality, vegetative growth and the development of size hierarchies. *J. Ecol.* **74**, 1069–1084.

Paul, N.D. and Ayres, P.G. (1987) Survival, growth and reproduction of groundsel (*Senecio vulgaris*) infected by rust (*Puccinia lagenophorae*) in the field during summer. *J. Ecol.* **75**, 61–71.

Paul, N.D. and Ayres, P.G. (1991) Changes in tissue freezing in *Senecio vulgaris* infected by rust (*Puccinia lagenophorae*). *Ann. Bot.* **68**, 129–133.

Ramsell, J. and Paul, N.D. (1990) Preferential grazing by molluscs of plants infected by rust fungi. *Oikos* **58**, 145–150.

Son, C.L. and Smith, S.E. (1988) Mycorrhizal growth responses: interactions between photon irradiance and phosphorus nutrition. *New Phytol.* **108**, 305–314.

Walters, D.R. (1981) *Root function in mildewed barley*. Ph.D. Thesis, University of Lancaster.

Walters, D.R. (1984) Shoot:root interrelationships: the effects of obligately biotrophic fungal pathogens. *Biol. Rev.* **60**, 47–70.

Wennstrom, A. and Ericson, L. (1991) Variation in disease incidence in grazed and ungrazed sites for the system *Pulsatilla pratensis–Puccinia pulsatillae*. *Oikos* **60**, 35–39.

Whipps, J.M. and Lewis, D.H. (1981) Patterns of translocation, storage and interconversion of carbohydrates. In: *Effects of Disease on the Physiology of the Growing Plant* (ed. P.G. Ayres). Cambridge University Press, Cambridge, pp. 47–83.

Whittaker, J.B. (1982) The effect of grazing by a chrysomelid beetle, *Gastrophysa viridula*, on the growth and survival of *Rumex crispus* on a shingle bank. *J. Ecol.* **70**, 291–296.

Zimmerman, M. and Pyke, G.H. (1988) Experimental manipulations of *Polemonium foliosissimum*: effects on subsequent nectar production, seed production and growth. *J.Ecol.* **76**, 777–789.

Zulu, J.N., Farrar, J.F. and Whitbread, R. (1991) Effects of phosphate supply on wheat seedlings infected with powdery mildew: carbohydrate metabolism of first leaves. *New Phytol.* **118**, 553–558.

10

Interaction of pests and pathogens through host predisposition

M.P. de Nooij, A. Biere and E.G.A. Linders

10.1 Introduction

10.1.1 *Interaction and predisposition*

Many plant species in agricultural and natural plant communities are hosts for a broad spectrum of parasites that overlap in their distribution in space and time. For example, there are over 60 well described diseases and an approximately equal number of known insect pests of rice (Bowling, 1980; Ou, 1980). The wild plant species *Plantago lanceolata* may interact with a wide range of viruses from at least 19 viral groups, two bacterial parasites, 29 plant pathogenic fungi, four parasitic nematodes and 15 phytophagous insects and molluscs (De Nooij and Mook, 1991). Insect herbivores may therefore frequently encounter a plant that has been previously attacked by a pathogen. Likewise, pathogens may often encounter a plant previously wounded or defoliated by a herbivore.

Alteration of the susceptibility of a plant to a pathogen or herbivore by non-genetic factors that act prior to the interaction between host and parasite is called predisposition (Yarwood, 1959; Schoeneweiss, 1975). Predisposition implies an effect on the host rather than on the parasite. Plants may be predisposed by abiotic stresses such as extremes of temperature, light or nutrients, drought, toxic substances or wounding, and by biotic stresses such as defoliation, sap removal or disease. Selection pressure on hosts through biotic stress is often assumed to be imposed by a single agent. This ignores the fact that plants are often host to multiple exploiters and that resource utilization is affected by the specific combination of exploiters. Results of several studies suggest that pathogens and herbivores may interact by altering the quality of their shared host plant (e.g. Karban *et al.*, 1987), both through alteration of allocation patterns and alteration of host defence strategies.

Here we consider such indirect interactions between herbivores and pathogens through predisposition of host plants and ask, how can a herbivore alter the suitability of its host plant (beneficially or detrimentally) for a subsequently invading pathogen and, conversely, how can a pathogen alter the suitability for a herbivore? Largely for the sake of convenience, we assume that the one type of exploiter attacks before the other. Coincident attack could, and does, occur but the interactions are more difficult to discern than if one precedes the other. Direct interactions between herbivores and pathogens, such as fungivory or vector mediation, and predisposition of host plants by pathogens to a second pathogen species, or by herbivores to a second herbivore species, are outside the scope of this chapter.

10.1.2 *Pathogens predisposing host plants of herbivores*

Plant pathogenic mediation of insect–host interactions is a frequently observed phenomenon (Jones, 1984; Hammond and Hardy, 1988). Diseased host plants represent a complex feeding niche for phytophagous insects, affecting reproductive biology and population dynamics in ways ranging from highly beneficial to acutely detrimental. The plant–pathogen complex may be viewed by the insects as an exploitable resource, representing a readily accessible source of refuge or nutrition, or an opportunity to colonize a formerly unsuitable host (Lewis, 1979; Raffa and Berryman, 1982; Steiner, 1984). For instance, larvae of the European corn borer, *Ostrinia nubilalis*, developed 20% faster on maize stalks infected with the fungus *Colletotrichum graminicola* than on uninoculated tissues. It was suggested that this stimulation was related to maceration of maize tissue and breakdown of its complex carbohydrates by fungal enzymes (Carruthers *et al.*, 1986). Likewise, Cobb *et al.* (1974) showed that infestation of ponderosa pine and white fir by bark beetles was strongly enhanced by the occurrence of fungal root pathogens. Nymphs of *Melanoplus differentialis*, a polyphagous grasshopper, were frequently observed feeding on sunflower leaves infected with the rust fungus *Puccinia helianthii*, whereas adjacent uninfected leaves were not attacked. Within a leaf, they fed on regions with the highest concentration of rust pustules (Lewis, 1979). Pathogens may also provide visual stimuli. Flight chamber experiments have demonstrated that alates of cereal aphids are attracted by the altered apparency (colour) of host leaves infected with barley yellow dwarf virus (BYDV) (Ajayi and Dewar, 1983).

Alternatively, changes in the diseased plant may render it less suitable or even lethal to herbivores due to reduced nutritional quality or high concentrations of toxic compounds (Isawa *et al.*, 1974; Mainer and Leath, 1978; Ahmad *et al.*, 1985; Potter, 1987). Effects attributable to direct pathogen–insect association are not always easily distinguished from the indirect interaction through host predisposition. Some pathogens are capable of utilizing the insect as an alternative host (Nakasuji and Kiritani, 1970; Granados and Meehan, 1975). Facultative mycophagous molluscs probably derive additional nutrition by consuming fungi from diseased hosts (Ramsell and Paul, 1990).

10.1.3 *Herbivores predisposing host plants of pathogens*

The most widely known example of predisposition of a host plant to pathogens is that of phytophagous insects breaking physical plant defences by providing wounds for those pathogens that cannot invade the host by themselves (e.g. De Nooij, 1988). Viruses, in particular, need such a predisposition before infection can take place (Carter, 1973; Harris and Maramorosch, 1980). However, the indirect impact of herbivores on pathogens via host plants is not necessarily beneficial to the pathogen. For example, the fungus *Verticillium dahliae* was less likely to cause symptoms of *Verticillium* wilt on cotton seedlings that had been previously exposed to spider mites than on unexposed cotton seedlings. Apparently, changes caused by herbivore attack reduced the suitability of the host plant to the pathogen (Karban *et al.*, 1987). Herbivory may also keep host abundance below levels at which fungal infections become epidemic, as suggested by Bowers and Sacchi (1991).

10.2 Response of herbivores and pathogens to altered allocation patterns in the host

Nutritional requirements of primary importance to most pathogens and phytophagous insects are normally met by healthy susceptible host plants. Alterations in nutrient quality, such as local depletion, conversion to a chemically inaccessible form, or retranslocation to distant organs or tissues within the plant may greatly influence insect and/or pathogen performance (Mattson, 1980).

10.2.1 *Defoliation*

The response of host plants to the drastic alteration of the root:shoot ratio due to defoliation by herbivorous insects may predispose them to pathogen attack. Frequent clipping predisposed red clover to infection by weakly parasitic species of *Fusarium*. Siddiqui *et al.* (1968) suggested that the increased susceptibility might be caused by the depletion of food reserves in the roots of the host. Similarly, Staley (1965) and Stephens (1971) observed that oaks suffered increased mortality after defoliation by insects early in the growing season. They suggested that defoliation resulted in the depletion of food reserves, causing loss of vigour and increased susceptibility to pathogens and other insects. Artificial defoliation of sugar maple trees resulted in higher mortality and a greater percentage of infection by *Armillaria mellea* (Wargo and Houston, 1973). Schoeneweiss (1975) investigated the effect of defoliation stress on disease susceptibility in woody stems of various tree species (Figure 10.1). Inoculated plants were exposed to complete artificial defoliation at weekly intervals for 8 weeks. In most cases stem cankers were initiated after 3 weeks of defoliation and canker incidence increased with the length of exposure to defoliation. However, in all these examples of predisposition following defoliation stress, it is difficult to distinguish between effects due to altered biomass (carbon, nutrient) allocation patterns, physical wounding, and induction of defence reactions, which will be discussed in the next sections.

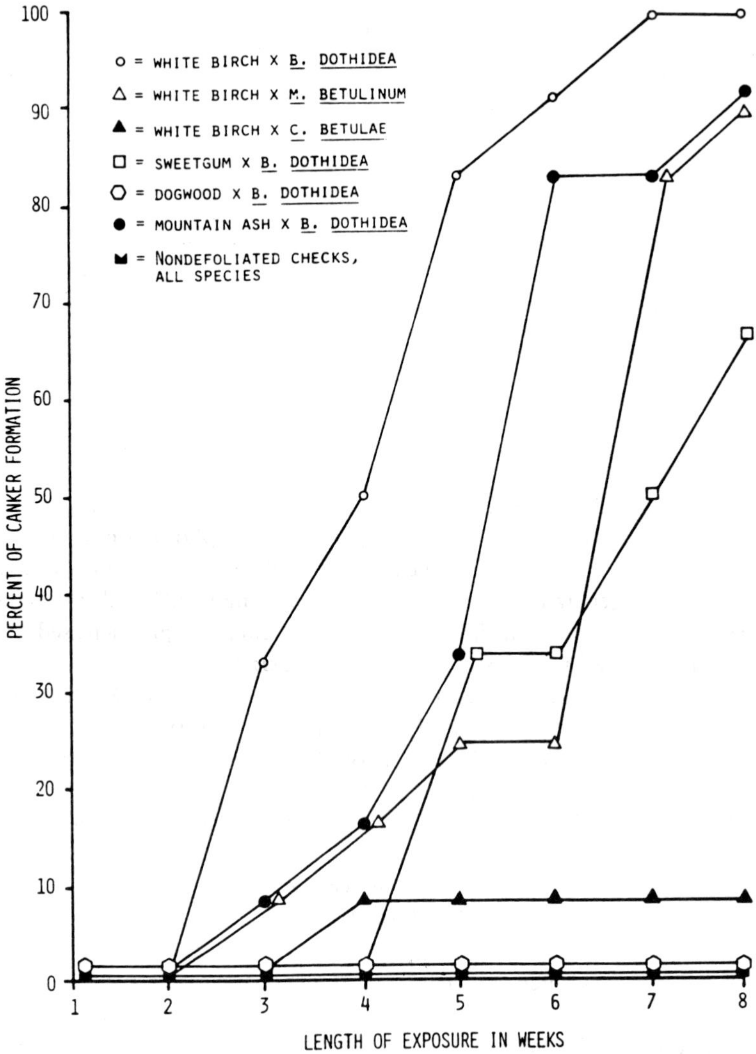

***Figure 10.1**. Effect of complete weekly defoliation on disease susceptibility in woody stems. Relation between length of exposure to defoliation stress and percentage of canker formation at inoculation points in 2-year-old, container-grown seedlings of European white birch (*Betula alba*) inoculated with* Botryosphaeria dothidea, Melanconium betulinum, *and* Cryptospora betulae; *and sweetgum (*Liquidambar styraciflua*), red-osier dogwood (*Cornus stolonifera*), and European mountain ash (*Sorbus aucuparia*) inoculated with* B. dothidea. *Percentage of canker formation is based on twelve replicate plants in each host–pathogen combination. Reproduced, with permission, from Schoeneweiss (1975), Annual Review of Phytopathology Vol. 13, © 1975 by Annual Reviews Inc.*

Table 10.1. *Mean fresh weight (mg) of leaf discs of* Tussilago farfara *(uninfected/infected by* Coleosporium tussilaginis*),* Poa pratensis *(uninfected/infected by* Puccinia poarum*) and* Rumex acetosa *(uninfected/infected by* Uromyces rumicis*), consumed by the slugs* Deroceras reticulatis *and* Arion hortensis (De Nooij, unpublished)

Slug sp.	*Tussilago*		*Poa*		*Rumex*	
	Uninf.	Inf.	Uninf.	Inf.	Uninf.	Inf.
Deroceras	6.4[b]	28.5[a]	0.20[a]	0.35[a]	3.30[b]	0.07[a]
Arion	22.4[a]	44.5[c]	0.39[a]	2.97[b]	4.01[b]	1.15[a]

For each treatment, $n = 10$. Values within a host species with different letters are significantly different at $P < 0.05$.

10.2.2 *Allocation of carbohydrates*

Infection of hosts by pathogenic micro-organisms often induces a net movement of carbon assimilates among plant organs. Movement of photoassimilates from non-infected leaves to infection sites may be enhanced, whereas export from infected leaves to other organs is generally reduced (see Chapter 8). Singh and Smalley (1969) showed that free sugar concentrations in the xylem sap of elms sharply increased after stem inoculation with the Dutch elm disease agent, *Ophiostoma ulmi*. The infected tissues might serve as an attractive energy reserve source for insects. On the other hand, notable reductions in soluble carbohydrates may be observed with systemic viral infection because of reductions in the rate of photosynthesis or due to changes in the metabolic pathway (Buchanan *et al.*, 1981).

Tussilago farfara infected with the rust *Coleosporium tussilaginis*, and *Poa pratensis* infected with the rust *Puccinia poarum*, were more palatable to slugs than uninfected leaf tissue (Table 10.1). Enhanced grazing occurred also on *Allium ursinum* leaf tissue infected with the rust *Puccinia sessilis* (Ramsell, 1989; Ramsell and Paul, 1990). Holligan *et al.* (1973), Hodges and Robinson (1977) and Ramsell (1989) showed that in the immediate area of rust infections, plant and fungal carbohydrates frequently accumulated, reaching as much as ten times the concentration found in healthy tissues. In addition to direct herbivore–pathogen interactions via mycophagy, the preference of herbivores for infected host plants may thus be mediated by host predisposition. In contrast to their feeding on healthy and rusted *T. farfara* and *P. pratensis*, slugs preferred healthy tissue of *Rumex acetosa* to that infected by rust (*Uromyces rumicis*) (Table 10.1). Thus, preferential grazing depends greatly on the specific combination of host, pathogen and herbivore.

10.2.3 *Allocation of nutrients*

There are many examples where pathogen activity directs nutrients to the site of infection or interferes with transport processes (Zimmerman and McDonough, 1978; Hancock and Huisman, 1981; Chapter 8). Like many stress factors, infection often results in an increase in the concentration of available nitrogen so that infected plant parts may become a better source of food for phytophagous insects (White, 1974). Herbivores feeding on such stressed plant tissues have a greater and more readily available supply of nitrogen in their food than they would have had if feeding on unstressed plants. However, this is not always beneficial. Piening (1972) found that substantial increases in nitrate-N in rusted rye were toxic to cattle. Furthermore, plant diseases may sometimes result in apparent nitrogen deficiency in the host, because the pathogen removes nitrogenous components from the plant for its own maintenance, growth and reproduction (Markkula and Laurema, 1964).

In a wide variety of plant species, infection with viruses commonly increases the concentration of nitrogen (and especially of free amino acids) in infected plant tissue compared to that of healthy tissues (Tu and Ford, 1970; Cooper and Selman, 1974; Matthews, 1981). Several studies reported that virus-infected tissues were preferentially grazed by herbivorous insects and contained higher levels of amino acids than did the less preferred tissues of healthy plants (Gildow, 1980; Ajayi and Dewar, 1982). The beneficial effects of plant virus infection on insect survival, growth or reproduction are well documented (Maramorosch, 1958; Jayarai and Seshadri, 1967; Gildow, 1983), but cases in which plant virus infection does not benefit phytophagous insects are known as well (e.g. Hodgeson, 1981; Khan and Saxena, 1985). Even for a single herbivore, the total effect may depend on the specific combination of pathogen and host. The green peach aphid, *Myzus persicae*, preferred sugarbeet leaves infected with beet western yellows virus (BWYV) and beet yellows virus (BYV) to symptomless leaves, but preferred symptomless leaves over leaves infected by curly top virus (CTV) (Macias and Mink, 1969).

Accumulation of nitrogenous compounds has also been found in infected tissue of plants infected by fungal pathogens (Reddy and Rao, 1976; Cherry *et al.*, 1978; Hodges and Robinson, 1977; Paul and Ayres, 1988; Roberts and Walters, 1988). In mildewed barley, nitrate concentrations decreased whereas the fraction of reduced nitrogenous compounds, including ammonium, proteins and free amino acids, was found to increase (Walters and Ayres, 1980). Also, in plants infected by nematodes (Meon *et al.*, 1978) and bacteria (Seitz and Hochter, 1964), accelerated breakdown of protein and concomitant liberation and translocation of amino acids may take place. The changes in nitrogen metabolism of diseased plants may vary substantially over time and with their growth conditions (Rhoades, 1985). For example, whereas pathogen protein may be made in addition to plant protein in nutrient-rich conditions, it may be made at the expense of plant protein if soil nitrogen is limiting plant growth (Matthews, 1981).

10.3 Responses of herbivores and pathogens to altered defence strategies of the host

10.3.1 *Constitutive defences*

Wounds. The first barrier that foliar pathogens and herbivores encounter is the cuticle and cell wall of epidermal cells. For all viruses and mycoplasmas, most bacteria, and many fungi, especially necrotrophs, wounding of plant tissue is a prerequisite for infection. Wounds can be generated by severe weather events, human activity, developmental processes of the plant, or activity of insects and other herbivores (Bostock and Stermer, 1989). Tissue wounding by herbivores and subsequent invasion by pathogens is frequently observed (e.g. Carter, 1973; Mills, 1983). An example of the synergistic interaction between a herbivorous insect and a fungus is the biological control of water hyacinth, *Eichhornia crassipes*, in the southern USA. Some 99% of the plants were killed within 6 weeks after the fungus *Cercospora rodmanii* was applied to plants affected by two weevil species, whereas neither the fungus nor the weevils were capable of controlling the water hyacinth on their own (Charudattan, 1986; Hasan and Ayres, 1990). Feeding and oviposition wounds as entry points for the fungus were thought to play a major role in this interaction.

Phyllosphere interactions. The phyllosphere provides a habitat for various saprophytic yeasts, black moulds, bacteria, and pathogenic micro-organisms that feed on nutrients diffusing from the leaf (Tukey, 1970), pollen, and aphid honeydew (Fokkema, 1971; Fokkema *et al.*, 1983). While phyllosphere interactions cannot be regarded as pure insect–pathogen interactions through host predisposition, changes in phyllosphere conditions through insect activities may considerably influence subsequent pathogenic infection. For instance, aphid honeydew enhances germination and superficial mycelial growth on wheat leaves, leading to more infection by the necrotrophic pathogen *Septoria nodorum* (Fokkema *et al.*, 1983; Dik, 1990), and enhances growth of sooty moulds resulting in inhibition of photosynthesis (e.g. Wood *et al.*, 1988; see also Chapter 1).

Secondary metabolites. Secondary plant metabolites can be defined as plant substances not essential for growth and reproductive metabolism (Vining, 1990). Many secondary metabolites can be regarded as defensive plant chemicals against herbivores and/or pathogens, but they may serve other biological functions as well (Levin, 1976; Jones, 1984; Vining, 1990; see also Chapter 4). Secondary metabolites are a diverse group of compounds that are synthesized via many different pathways and which include, for example, tannins, alkaloids, phenolics, terpenoids, lectins, lignin, callose, and various compounds serving as enzyme inhibitors and phytoalexins. Secondary metabolites are not homogeneously distributed in plants. For instance, young leaves generally contain higher levels of secondary metabolites than senescent leaves (McKey, 1979).

Specialist insects have evolved many different ways in which they can cope with plant defence chemicals. Some herbivorous insects have evolved detoxifying mechanisms or preferentially attack plants or plant parts with low levels of secondary metabolites (Rosenthal and Janzen, 1979; Rhoades, 1985). Another strategy involves a mutualistic relationship with gut micro-organisms which can detoxify secondary metabolites (Levin, 1976; Rosenthal, 1983), and which may formerly have been associated with plants (Pirozynski and Hawksworth, 1988). In ambrosia galls, the fungi (*Botryosphaeria* spp. and related anamorphs) convert plant tissue into a highly nutritive and less toxic food source for the developing gall midge larvae (Bissett and Borkent, 1988). It is suggested that the rapid adaptation to phylogenetically diverse plant taxa by these ambrosia gall midges was facilitated by the symbiosis with the fungus.

In some plant species, the chemical compounds that deter herbivores are derived from a mutualistic relationship with a fungus. Several studies have supported the hypothesis that hosts infected with fungal endophytes suffer less herbivore damage than non-infected hosts (reviewed by Siegel *et al.*, 1987; Carroll, 1988; Clay, 1988a,b, 1990). As many grasses are remarkably free of defensive chemicals, the endophytic fungi seem to fill in a 'vacant niche' with their ability to produce alkaloids that are effective against insect and mammalian herbivores (Siegel *et al.*, 1987; Clay, 1988a). Grasses infected with seedborne *Acremonium* endophytic fungi produce more inflorescences and seeds, have a greater vegetative vigour, and suffer less herbivore damage than uninfected plants (Clay, 1990). In this case, the endophyte can be regarded as an analogue of a 'maternally inherited defence gene' that regulates the production of alkaloids, and that has a 'pleiotropic effect', i.e. it increases host vigour, factors which both contribute to host fitness.

In general, the role of pathogen-derived toxic compounds in host defence is less well studied than that of host-derived secondary metabolites elicited by pathogens (induced resistance) and these are not always easily distinguished. Dowd (1989) showed that, on artificial medium, fusaric acid from *Fusarium* spp. had a synergistic effect on gossypol and 6-methoxy-2-benzoxazolinone (two secondary compounds of cotton) with respect to their toxicity to larvae of the corn earworm, *Heliothis zea*. Alfalfa leaves infected with *Verticillium* had lower digestibility for the southern army worm, *Spodoptera eridania* (Kingsley *et al.*, 1983). *Verticillium*-infected leaves of cotton were found to be more resistant against spider mites than non-infected plants (Karban *et al.*, 1987). Although all these phenomena were attributed to induced production of secondary compounds by the host plant, in both *Verticillium* examples fungal metabolites could have contributed to resistance (Wicklow, 1988).

10.3.2 *Induced resistance*

Secondary metabolites are not only constitutively present in plant tissues, they can also be locally or systemically induced. The induced compound is not necessarily newly synthesized. Induction may involve an increase in the amount of a pre-existing secondary metabolite. The response is usually quantitative and generally depends on the nutritional condition of the plant and on the severity of the inducing event. Many aspects of inducible defence have been reviewed extensively, such as the types of secondary metabolites involved (Levin, 1976), defence-related proteins (Bowles, 1990; Ryan, 1990), molecular interactions (Dixon and Lamb, 1990), their induction by herbivores (Rhoades, 1985; Karban and Myers, 1989; Larsson, 1989), by pathogens (Kuc, 1982; Sequeira, 1983; De Wit, 1985) and by viruses (Bol *et al.*, 1990), wound healing (Bostock and Stermer, 1989), and ecological and evolutionary aspects (Harvell, 1990). These reviews clearly show that there is a high degree of variability in the type and rate of synthesis of allelo-chemicals induced by a single parasite in different hosts, and that the same is true for induction by different parasites in a common host.

An increasing number of studies show that resistance in host plants to pathogens can be induced by a previous herbivore attack, and vice versa. An example of the occurrence of pathogen-induced resistance against herbivores is the systemic resistance of a hypersensitive tobacco cultivar to the green peach aphid, *Myzus persicae*, following inoculation with tobacco mosaic virus (TMV). Resistance was also induced against TMV itself, two fungi and a bacterial pathogen (McIntyre *et al.*, 1981). In the same plant–pathogen system, larvae of *Manduca sexta*, the tobacco hornworm, showed reduced growth on tobacco leaves that were locally and systemically infected with TMV (Hare, 1983). The factors responsible for the induced resistance were not determined in either of these studies, but both illustrate that a single pathogen can induce protection against a broad range of organisms. By contrast, Apriyanto and Potter (1990) found that infection with tobacco necrosis virus (TNV) resulted in systemic protection of cucumber against the anthracnose fungus *Colletotrichum lagenarium*, but not against three generalist and one specialist herbivores. Challenge with TNV did not result in a systemic increase in cucurbitacins. Cucurbitacins can be systemically induced by herbivore damage and are known to deter non-adapted herbivores, but have no known effect on pathogens (Tallamy and Krischik, 1989). It was concluded therefore that the mechanisms and specificity of resistance induced in cucurbits by pathogens are different from those induced by herbivores (Apriyanto and Potter, 1990).

Parasites often have multiple effects on their hosts. This makes it difficult to separate the effects of a specific, induced metabolite from effects of, for instance, altered nutritional status or from effects of other induced compounds. Fischer *et al.* (1990) circumvented this problem by isolating the phytoalexin glyceollin from soybean challenged with *Phytophthora megasperma* f.sp. *glycinea*, a fungus inhibited by glyceollin, and subsequently spraying the isolated glyceollin on leaves of the common bean, *Phaseolus vulgaris*, a species that does not produce glyceollin. The sprayed bean leaves were then fed in a preference test to three different herbivores, one polyphagous species, one species marginally found on soybean, and one adapted species. Only the adapted bean leaf beetle was not deterred from feeding, indicating a broad biological action spectrum for glyceollin.

A number of inducible secondary metabolites are now known to affect a wide range of plant parasites, including both herbivores and pathogens. Krischik *et al.* (1991) showed that the inducible alkaloid, nicotine, incorporated in various concentrations on artificial media, had almost no effect on larval growth and survival of *Manduca sexta*, a specialist herbivore commonly found on tobacco, but strongly inhibited a generalist herbivore, *Trichoplusia ni*, and five *Pseudomonas* species reared on the same media. The flavonoid, rutin, on the other hand, had only a slight inhibitory effect on the generalist herbivore, *T. ni*. These examples clearly show that interactions between pathogens and herbivores, through host predisposition in which secondary metabolites are involved, can be highly species-specific.

10.4 Response of herbivores and pathogens to altered host density and community structure

We have so far considered predisposition at the level of individual host plants, but the apparency of host plants and their suitability for a pathogen or herbivore is also affected by factors at the population and community level (Price *et al.*, 1980; Thresh, 1981). Pathogens may affect host population density and distribution, and hence the distribution of herbivorous insects, as in the case of *Scolytus* beetles and Dutch elm disease (Barnes, 1976). Likewise, herbivores may drastically alter patterns of plant succession and affect host population density and longevity (Brown, 1982), and alter the opportunity for pathogens to become epidemic.

Bowers and Sacchi (1991) found that the percentage of leaflets of purple clover, *Trifolium pratense*, infected with rust fungus, *Uromyces trifolii*, was four times lower in grazed (control) plots than in deer exclosures. They suggested that this difference was due to the 40% higher relative abundance of *T. pratense* in exclosures, which enabled a more rapid increase in the pathogen population. Increases in the rate of pathogen spread at higher host densities are well documented (e.g. Burdon and Chilvers, 1982). Herbivore activity may thus act as a 'constraint to epidemic spread' of the pathogen, not only by (selective) grazing of infected leaf area of individual hosts (Barbe, 1964; Lewis, 1979; Steiner, 1984; Ericson and Wennström (in Carlsson *et al.*, 1990); Ramsell and Paul, 1990) resulting in a reduction in inoculum, but also by reducing the density of suitable hosts or alteration of the vegetation composition.

A negative relationship between herbivore presence and disease incidence does not necessarily result from host predisposition. For instance, Wennström and Ericson (1991) found that the incidence of the host-sterilizing, systemic rust, *Puccinia pulsatillae*, on *Pulsatilla pratensis* was more than ten-fold lower in grazed than in ungrazed localities. They suggested that, since plants infected with *P. pulsatilla* have increased vegetative vigour, infected hosts would be better able to compete with grasses in the dense swards of ungrazed localities, and would become relatively more abundant in these sites. In addition, recruitment of healthy hosts in ungrazed localities was low because establishment of seedlings, which are uninfected, appeared to be a rare event in ungrazed, closed swards. Thus, in *P. pratensis* populations, herbivory would merely alter the magnitude of the competitive advantage induced by rust infection, and there would be no direct predisposition. This is in contrast to the effect of many endophytic fungi (e.g. Clay, 1990) which not only increase the vegetative vigour of their hosts but, in addition, directly predispose them against herbivory.

10.5 Concluding remarks

Direction of host predisposition: outcome of a multi-dimensional process. We have seen that alteration of host plant quality following pathogen attack can enhance or reduce the suitability of a host for subsequently invading herbivores, and that herbivores can exert similar effects on pathogens. Enhanced suitability can be mediated through increased tissue concentrations of amino acids, lipids or non-structural carbohydrates, but other mechanisms such as tissue maceration, breakdown of secondary metabolites, creation of entry sites, and changes in visual appearance may also be important (Table 11.2). Most common among examples of reduced suitability through host predisposition are cases involving various classes of induced and/or pathogen-derived allelo-chemicals, although some studies report reduced suitability because of decreased levels of crude protein or carbohydrates or, occasionally, the presence of toxic nutrient concentrations. Altered suitability is a broad term that can include differences in various aspects of performance at any stage of the lifecycle. The consequences of altered suitability, when measured in terms of fitness of the ensuing pathogen or herbivore, are not always clear. For instance, larvae of the southern armyworm, *Spodoptera eridania*, appear to be able to compensate for their lower conversion efficiency of alfalfa infected with *Verticillium* by increased consumption rates, so that their relative growth rate is unaffected (Kingsley *et al.*, 1983).

Generalizations about the direction in which the suitability of hosts for ensuing invaders is changed are hampered in several ways. First, the impact of alterations in a single host attribute varies strongly among herbivore or pathogen species encountering the predisposed host. Secondary metabolites deterring a generalist herbivore may be detoxified, tolerated, or sequestered by a specialist herbivore that may even use them as a feeding or oviposition cue (e.g. Rhoades, 1985). Second, host alteration and its impact on ensuing parasites may also depend on the biotic or abiotic environment and the developmental stage of both host and parasite. For instance, induced resistance of cotton to *Verticillium albo-atrum* decreased with increasing host density, possibly as a consequence of the decreased availability of resources (Karban *et al.*, 1989). Third, a single pathogen or herbivore may simultaneously affect several host attributes. The direction in which host suitability is altered is therefore the net result of a complex interaction with this specific organism.

Knowledge of changes in a single host attribute may have little predictive value for altered suitability even for a given pathogen–host–herbivore system. This is nicely illustrated by the steroidal glycoalkaloid, α-tomatine, which deters insect herbivores such as the Colorado potato beetle, *Leptinotarsa decemlineata*, from feeding on tomato plants and has been implicated in the resistance of wild *Lycopersicon* species to this herbivore (Sinden *et al.*, 1978). Infection of tomato with TMV resulted in a 45% increase in tomatine content, but simultaneously increased the total N concentration by 46% (Hare and Dodds, 1987). Survival of *L. decemlineata* larvae increased linearly with N concentration, whereas tomatine level could not explain any additional variation in survival (Hare and Dodds, 1987). Interactions between allelo-chemicals may also affect the direction in which suitability is altered. Ma (1972) showed that larvae of *Pieris brassicae* were repelled from feeding by α-tomatine but that, in the presence of the glucosinolate, sinigrin, low concentrations of α-tomatine actually enhanced feeding by *P. brassicae*. Sinigrin, an allelo-chemical characteristic of Cruciferae, is converted to allyl isothiocyanate which is toxic both to fungi (e.g. *Peronospora*) and many herbivores not specialized for feeding on Cruciferae, e.g. *Papilio polyxenes*, but is used as a feeding and oviposition stimulant by *P. brassicae*.

Effects of predisposition on host fitness. As pointed out by Karban and Myers (1989), altered host quality does not necessarily affect herbivore or pathogen preference or

Table 10.2. *Examples of host plant characteristics altered by pathogens and affecting their suitability for herbivores, or vice versa*

Host characteristic	Specific alteration by herbivore/pathogen	Predisposition: by herbivore (H) or pathogen (P)	Suitability: direction of change[1]	Ref[2]
Plant population				
Host availability	Absolute or relative host density or abundance	H,P	+ or –	6,31,38
Host apparency	Olfactorial or visual appearance (necrosis/chlorosis, stunting, etc.)	P	+ or –	3
Opportunity to (re-)colonize host	Presence of alternative hosts for completion of life cycle or perennation	H,P	+ or –	37
Individual plant				
Tissue accessibility	Oviposition or feeding wounds	H	+	8,9,13,14,28
Tissue exploitability	Tissue maceration or breakdown	H,P	+	7,22,32,37
Tissue availability	Phenology, tissue removal, biomass allocation patterns, growth rate (e.g. reduction of photosynthetic active tissue, inhibition of xylem transport)	H,P	+ or –	34,36
Tissue quality:				
(a) Nutritional status	Formation/depletion of local sinks of C- and/or N-metabolites, C/N balance, concentration of metabolites (e.g. due to fungus-induced wilting), short/long term alteration of nutrient levels (e.g. after defoliation)	H,P	+ or –	1,2,11,15,16, 18,23,24,25, 26,29,30,41
(b) Composition of defence chemicals	Local/systemic levels of secondary compounds, growth regulators or proteins (e.g. terpenoids, protease inhibitors and pathogenesis-related proteins)	H,P	+ or –	4,5,10,12,17, 19,20,21,27, 33,35,38

[1] Direction of predisposition: beneficial (+) or detrimental (–) to ensuing herbivore or pathogen.

[2] References: 1. Ahmad *et al.*, 1985; 2. Ajayi and Dewar, 1982; 3. Ajayi and Dewar, 1983; 4. Apriyanto and Potter, 1990; 5. Bol *et al.*, 1990; 6. Bowers and Sacchi, 1991; 7. Carruthers *et al.*, 1986; 8. Carter, 1973; 9. De Nooij, 1988; 10. De Wit, 1985; 11. Gildow, 1980; 12. Hare, 1983; 13. Harris and Maramorosch, 1980; 14. Hasan and Ayres, 1990; 15. Hodgeson, 1981; 16. Isawa *et al.*, 1974; 17. Karban and Myers, 1989; 18. Khan and Saxena, 1985; 19. Krischik *et al.*, 1991; 20. Kuc, 1982; 21. Larsson, 1989; 22. Lewis, 1979; 23. Macias and Mink, 1969; 24. Mainer and Leath, 1978; 25. Markkula and Laurema, 1964; 26. Mattson, 1980; 27. McIntyre *et al.*, 1981; 28. Mills, 1983; 29. Piening, 1972; 30. Potter, 1987; 31. Price *et al.*, 1980; 32. Raffa and Berryman, 1982; 33. Rhoades, 1985; 34. Schoeneweiss, 1975; 35. Sequeira, 1983; 36. Siddiqui *et al.*, 1968; 37. Steiner, 1984; 38. Tallamy and Krischik, 1989; 39. Thresh, 1981; 40. Wennström and Ericson, 1991; 41. White, 1974.

performance, and altered pathogen or herbivore performance does not necessarily affect plant fitness. In principle, however, pathogens that alter the apparency, suitability or quality of the host for herbivores that attack later, or vice versa, may significantly affect the fitness of the host. Clear examples are the enhanced persistence of endophyte-infected tall fescue in the presence of mammalian herbivores (e.g. Clay, 1990), and the reduced survival of weevil-damaged water hyacinth invaded by the pathogenic fungus *Cercospora rodmanii* (Hasan and Ayres, 1990). The examples have practical applications in crop protection and weed control, respectively. Obviously, the effect of pathogen–herbivore interactions, via host predisposition, on the fitness of the host will be more complicated than this in many natural plant communities. Plants are usually hosts to a large number of herbivore and pathogen species and the net effect of predisposition on the host will be a function of all these interactions. Nevertheless, such examples show that indirect interactions between micro-organisms and herbivores mediated by the plant add to the complexity of interactions that can significantly affect the fitness of plants both in natural populations and in managed systems.

Acknowledgements

We would like to thank Jos van Damme, Nigel Paul, Jacqui Shykoff and Jan Woldendorp for their comments on an earlier draft of the manuscript.

References

Ahmad, S., Govindarajan, S., Funk, C.R. and Johnson-Cicalese, J.M. (1985) Fatality of house crickets on perennial ryegrasses infected with a fungal endophyte. *Entomol. Exp. Appl.* **39**, 183–190.

Ajayi, O. and Dewar, A.M. (1982) The effect of barley yellow dwarf virus on honeydew production by the cereal aphids *Sitobion avenae* and *Metopolophium dirhodum*. *Ann. Appl. Biol.* **100**, 203–212.

Ajayi, O. and Dewar, A.M. (1983) The effect of barley yellow dwarf virus on field populations of the cereal aphids, *Sitobion avenae* and *Metopolophium dirhodum*. *Ann. Appl. Biol.* **103**, 1–11.

Apriyanto, D. and Potter, D.A. (1990) Pathogen-activated induced resistance of cucumber: response of arthropod herbivores to systemically protected leaves. *Oecologia* **85**, 25–31.

Barbe, G.D. (1964) The relation of European earwig to snapdragon rust. *Phytopathol.* **54**, 369–370.

Barnes, B.U. (1976) Succession in deciduous swamp communities of southeastern Michigan formerly dominated by American elm. *Can. J. Bot.* **54**, 20–24.

Bissett, J. and Borkent, A. (1988) Ambrosia galls: The significance of fungal nutrition in the evolution of the Cecidomyiidae (Diptera). In: *Coevolution of Fungi with Plants and Animals* (eds K.A. Pirozynski and D.L. Hawksworth). Academic Press, London, pp. 203–225.

Bol, J.F., Linthorst, H.J.M. and Cornelissen, B.J.C. (1990) Plant pathogenesis-related proteins induced by virus infection. *Ann. Rev. Phytopathol.* **28**, 113–138.

Bostock, R.M. and Stermer, B.A. (1989) Perspectives on wound healing in resistance to pathogens. *Ann. Rev. Phytopathol.* **27**, 343–371.

Bowers, M.A. and Sacchi, C.F. (1991) Fungal mediation of a plant-herbivore interaction in an early successional plant community. *Ecology* **72**, 1032–1037.

Bowles, D.J. (1990) Defense-related proteins in higher plants. *Ann. Rev. Biochem.* **59**, 873–907.

Bowling, C.C. (1980) Insect pests of the rice plant. In: *Rice: Production and Utilization* (ed. B.S. Luh). AVI, Westport CN, pp. 260–288.

Brown, V.K. (1982) The phytophagous insect community and its impact on early successional habitats. In: *Proceedings of the 5th International Symposium on Insect-Plant Relationships* (eds J.H. Visser and A.K. Minks). Pudoc, Wageningen, pp. 205–213.

Buchanan, B.B., Hutcheson, S.W., Magyarosi, A.C. and Montalbini, P. (1981) Photosynthesis in healthy and diseased plants. In: *Effects of Disease on the Physiology of the Growing Plant.* (ed. P.G. Ayres). Cambridge University Press, Cambridge, pp. 13–28.

Burdon, J.J. and Chilvers, G.A. (1982) Host density as a factor in plant disease ecology. *Ann. Rev. Phytopathol.* **20**, 143–166.

Carlsson, U., Elmqvist, T., Wennström, A. and Ericson, L. (1990) Infection by pathogens and population age of host plants. *J. Ecol.* **78**, 1094–1105.

Carrol, G. (1988) Fungal endophytes in stems and leaves: from latent pathogen to mutualistic symbiont. *Ecology* **69**, 2–9.

Carruthers, R.I., Bergstrom, G.C. and Haynes, P.A. (1986) Accelerated development of the European corn borer, *Ostrinia nubilalis* (Lepidoptera: Pyralidae), induced by interactions with *Colletotrichum graminicola* (Melanconiales: Melanconiaceae), the causal fungus of maize anthracnose. *Ann. Entomol. Soc. Am.* **79**, 385–389.

Carter, W. (1973) *Insects in Relation to Plant Disease.* John Wiley and Sons, New York.

Charudattan, R. (1986) Integrated control of water hyacinth (*Eichhornia crassipes*) with a pathogen, insects and herbicides. *Weed Sci.* **34** (suppl. 1), 26–30.

Cherry, J.P., Beuchat, L.R. and Koehler, P.E. (1978) Soluble proteins and enzymes as indicators of change in peanuts infected with *Aspergillus flavus*. *J. Agric. Food Chem.* **26**, 242–245.

Clay, K. (1988a) Clavicipitaceous fungal endophytes of grasses: coevolution and the change from parasitism to mutualism. In: *Coevolution of Fungi with Plants and Animals* (eds K.S. Pirozynski and D.L. Hawksworth). Academic Press, London, pp. 79–105.

Clay, K. (1988b) Fungal endophytes of grasses: a defensive mutualism between plants and fungi. *Ecology* **69**, 10–16.

Clay, K. (1990) Fungal endophytes of grasses. *Ann. Rev. Ecol. Syst.* **21**, 275–297.

Cobb, F.W., Parmeter, J.R., Wood, D.L. and Stark, R.W. (1974) Root pathogens as agents predisposing ponderosa pine and white fir to bark beetles. *Proceedings 4th International Conference on* Fomes annosus, Athens, Georgia (1973), Forest Service USDA, Asheville NC, pp. 8–15.

Cooper, P. and Selman, J.W. (1974) An analysis of the effects of tobacco mosaic virus on growth and the changes in the free amino compounds in young tomato plants. *Ann. Bot.* **38**, 625–638.

De Nooij, M.P. (1988) The role of weevils in the infection process of the fungus *Phomopsis subordinaria* in *Plantago lanceolata*. *Oikos* **52**, 51–58.

De Nooij, M.P. and Mook, H.J. (1991) Interactions with organisms other than plants. In: *Plantago, a Multidisciplinary Study, Ecological Studies*, Volume 89 (eds P.J.C. Kuiper and M. Bos). Springer-Verlag, Heidelberg, pp. 52–68.

De Wit, P.J.G.M. (1985) Induced resistance to fungal and bacterial diseases. In: *Mechanisms of Induced Resistance to Plant Diseases* (ed. R.S.S. Fraser). Martinus Nijhoff/W.Junk, Boston, pp. 405–424.

Dik, A.J. (1990) *Population dynamics of phyllosphere yeasts: influence of yeasts on aphid damage, diseases and fungicide activity in wheat.* Ph.D. Thesis, University of Utrecht, The Netherlands.

Dixon, R.A. and Lamb, C.J. (1990) Molecular communication in interactions between plants and microbial pathogens. *Ann. Rev. Physiol. Plant Mol. Biol.* **41**, 339–367.

Dowd, P.F. (1989) Fusaric acid. A secondary fungal metabolite that synergizes toxicity of cooccurring host allelochemicals to the corn earworm, *Heliothis zea* (Lepidoptera). *J. Chem. Ecol.* **15**, 249–254.

Fischer, D.C., Kogan, M. and Paxton, J. (1990) Effect of glyceollin, a soybean phytoalexin, on feeding by three phytophagous beetles (Coleoptera: Coccinellidae and Chrysomelidae): dose versus response. *Environ. Entomol.* **19**, 1278–1282.

Fokkema, N.J. (1971) The effect of pollen in the phyllosphere of rye on colonization by saprophytic fungi and on infection by *Helminthosporium sativum* and other leaf pathogens. *Neth. J. Plant Pathol.* **77** (suppl.1), 1–60.

Fokkema, N.J., Riphagen, I., Poot, R.J. and de Jong, C. (1983) Aphid honeydew, a potential stimulant of *Cochliobolus sativus* and *Septoria nodorum* and the competitive role of saprophytic mycoflora. *Trans. Br. Mycol. Soc.* **81**, 355–363.

Gildow, F.E. (1980) Increased production of alatae by aphids reared on oats infected with the yellow dwarf virus. *Ann. Entomol. Soc. Am.* **73**, 343–347.

Gildow, F.E. (1983) Influence of barley yellow dwarf virus-infected oats and barley on morphology of aphid vectors. *Phytopathol.* **73**, 1196–1199.

Granados, R.R. and Meehan, D.J. (1975) Pathogenicity of the corn stunt agent to an insect vector, *Dalbulus elimatus*. *J. Invert. Pathol.* **26**, 313–320.

Hammond, A.M. and Hardy, T.N. (1988) Quality of diseased plants as hosts for insects. In: *Plant Stress-Insect Interactions* (ed. E.A. Heinrichs). John Wiley and Sons, New York, pp. 381–432.

Hancock, J.G. and Huisman, O.C. (1981) Nutrient movement in host-pathogen systems. *Ann. Rev. Phytopathol.* **19**, 309–331

Hare, J.D. (1983) Manipulation of host suitability for herbivore pest management. In: *Variable Plants and Herbivores in Natural and Managed Systems* (eds R.F. Denno and M.S. McClure). Academic Press, New York, pp. 655–680.

Hare, J.D. and Dodds, J.A. (1987) Survival of the Colorado potato beetle on virus infected tomato in relation to plant nitrogen and alkaloid content. *Entomol. Exp. Appl.* **44**, 31–35.

Harris, K.F. and Maramorosch, K. (1980) *Vectors of Plant Pathogens*. Academic Press, New York.

Harvell, C.D. (1990) The ecology and evolution of inducible defenses. *Quart. Rev. Biol.* **65**, 323–340.

Hasan, S. and Ayres, P.G. (1990) The control of weeds through fungi: principles and prospects. *New Phytol.* **115**, 201–222.

Hodges, C.F. and Robinson, P.W. (1977) Sugar and amino acid content of *Poa pratensis* infected with *Ustilago striiformis* and *Urocystis agropyri*. *Physiol. Plant.* **41**, 25–28.

Hodgeson, C.J. (1981) Effects of infection with the cabbage black ringspot strain of turnip mosaic virus on turnip as a host to *Myzus persicae* and *Brevicoryne brassicae*. *Ann. Appl. Biol.* **98**, 1–14.

Holligan, P.M., Chen, C. and Lewis, D.H. (1973) Changes in the carbohydrate composition of leaves of *Tussilago farfara* during infection by *Puccinia poarum*. *New Phytol.* **72**, 947–955.

Isawa, K., Abe, A. and Nishihara, N. (1974) Influence of diseases on the chemical composition and nutritive value of forage crops. I. Chemical composition of Italian ryegrass infected with crown rust. *Ann. Phytopath. Soc. Jpn.* **40**, 86–90.

Jayarai, S. and Seshadri, A.R. (1967) Preference of the leaf hopper *Empoasca kerri* Pruthi (Homoptera: Jassidae) to pigeon pea (*Cajanus cajan* (l.) Millsp.) plants infected with sterility mosaic virus. *Current Sci.* **36**, 353–355.

Jones, C.G. (1984) Microorganisms as mediators of plant resource exploitation by insect herbivores. In: *A New Ecology* (eds P.W. Price, C.N. Slobodchikoff and W.S. Gaud). John Wiley and Sons, New York, pp. 53–99.

Karban, R., Adamchak, R. and Schnathorst, W.C. (1987) Induced resistance and interspecific competition between spider mites and a vascular wilt fungus. *Science* **235**, 678–680.

Karban, R., Brody, A.K. and Schnathorst, W.C. (1989) Crowding and a plant's ability to defend itself against herbivores and diseases. *Am. Nat.* **134**, 749–760.

Karban, R. and Myers, J.H. (1989) Induced plant responses to herbivory. *Ann. Rev. Ecol. Syst.* **20**, 331–348.

Khan, Z.R. and Saxena, R.C. (1985) Behavior and biology of *Nephotettix virescens* (Homopterea: Cicadellidae) on tungro virus-infected rice plants: Epidemiology implications. *Environ. Entomol.* **14**, 297–304.

Kingsley, P., Scriber, J.M., Grau, C.R. and Delwiche, P.A. (1983) Feeding and growth performance of *Spodoptera eridania* (Noctuidae: Lepidoptera) on 'vernal' alfalfa, as influenced by *Verticillium* wilt. *Prot. Ecol.* **5**, 127–134.

Krischik, V.A., Goth, R.W. and Barbosa, P. (1991) Generalized plant defense: effects on multiple species. *Oecologia* **85**, 562–571.

Kuc, J. (1982) Induced immunity to plant disease. *BioScience* **32**, 845–860.

Larsson, S. (1989) Stressful times for the plant stress – insect performance hypothesis. *Oikos* **56**, 277–283.

Levin, D.A. (1976) The chemical defense of plants to pathogens and herbivores. *Ann. Rev. Ecol. Syst.* **7**, 121–159.

Lewis, A.C. (1979) Feeding preference for diseased and wilted sunflower in the grasshopper *Melanoplus differentialis. Entomol. Exp. Appl.* **26**, 202–207.

Ma, W.G. (1972) Dynamics of feeding responses in *Pieris brassica* as a function of chemosensory input: A behavioral, ultrastructural and electrophysiological study. *Meded. Landbouwhogesch. Wageningen* **72**, 162.

Macias, W. and Mink, G.I. (1969) Preference of green peach aphids for virus-infected sugarbeet leaves. *J. Econ. Entomol.* **62**, 28–29.

Mainer, A. and Leath, K.T. (1978) Foliar diseases alter carbohydrate and protein levels in leaves of alfalfa and orchard grass. *Phytopathol.* **68**, 1252–1255.

Maramorosch, K. (1958) Beneficial effects of virus-diseased plants on non-vector insects. *Tijdschr. Plantenz.* **64**, 383–391.

Markkula, M. and Laurema, S. (1964) Changes in the concentration of free amino acids in plants induced by virus diseases and the reproduction of aphids. *Ann. Agric. Fenn.* **3**, 265–271.

Matthews, R.E.F. (1981) *Plant Virology*. Academic Press, New York.

Mattson, W.J. (1980) Herbivory in relation to plant nitrogen content. *Ann. Rev. Ecol. Syst.* **11**, 119–161.

McIntyre, J.L., Dodds, J.A. and Hare, J.D. (1981) Effects of localized infections of *Nicotiana tabacum* by tobacco mosaic virus on systemic resistance against diverse pathogens and an insect. *Phytopathol.* **71**, 297–301.

McKey, D. (1979) The distribution of secondary compounds within plants. In: *Herbivores: Their Interaction with Secondary Plant Metabolites* (eds G.A. Rosenthal and D.H. Janzen). Academic Press, New York, pp. 55–133.

Meon, S., Fisher, J.M. and Wallace, H.R. (1978) Change in free proline following infection of plants with either *Meloidogyne javanica* or *Agrobacterium tumefaciens. Physiol. Plant Pathol.* **12**, 251–256.

Mills, J.T. (1983) Insect-fungus associations influencing seed deterioration. *Phytopathol.* 73, 330–335.

Nakasuji, F. and Kiritani, K. (1970) Ill-effects of rice dwarf virus upon its vector *Nephotettix cincticeps* Uhler (Hemiptera: Deltocephalidae) and its significance for changes in relative abundance of infected individuals among vector populations. *Appl. Entomol. Zool.* **5**, 1–12.

Ou, S.H. (1980) Rice plant diseases. In: *Rice: Production and Utilization* (ed. B.S. Luh). AVI Westport, CN, pp. 235–259.

Paul, N.D. and Ayres, P.G. (1988) Nutrient relations of groundsel (*Senecio vulgaris*) infected by rust (*Puccinia lagenophorae*) at a range of nutrient concentrations I. Concentrations, contents and distribution of N, P and K. *Ann. Bot.* **61**, 489–498.

Piening, L.G. (1972) Effects of leaf rust on nitrate in rye. *Can. J. Plant Sci.* **52**, 842–843.

Pirozynski, K.A. and Hawksworth, D.L. (1988) Coevolution of fungi with plants and animals: Introduction and overview. In: *Coevolution of Fungi with Plants and Animals* (eds K.A. Pirozynski and D.L. Hawksworth). Academic Press, London, pp. 1–29.

Potter, L.R. (1987) Effect of crown rust on regrowth, competitive ability and nutritional quality of perennial ryegrass and Italian ryegrass. *Plant Pathol.* **36**, 455–461.

Price, P.W., Bouton, C.E., Gross, P., McPheron, B.A., Thompson, J.N. and Weis, A.E. (1980) Interactions among three trophic levels: influence of plants on interactions between insect herbivores and natural enemies. *Ann. Rev. Ecol. Syst.* **11**, 41–65.

Raffa, K.F. and Berryman A.A. (1982) Physiological differences between lodgepole pines resistant and susceptible to mountain pine beetle and associated microorganisms. *Environ. Entomol.* **11**, 486–492.

Ramsell, J. (1989) *The influence of plant quality on invertebrate herbivory*. Ph.D. Thesis, University Lancaster, UK.

Ramsell, J. and Paul, N.D. (1990) Preferential grazing by molluscs of plants infected by rust fungi. *Oikos* **58**, 145–150.

Reddy, M.N. and Rao, A.S. (1976) Changes in the composition of free and protein amino acids in groundnut leaves induced by infection with *Puccinia arachidis* Speg. *Acta Phytopathol.* **11**, 167–172.

Rhoades, D.F. (1985) Offensive-defensive interactions between herbivores and plants: their relevance in herbivore population dynamics and ecological theory. *Am. Nat.* **125**, 205–238.

Roberts, A.M. and Walters, D.R. (1988). Nitrogen assimilation and metabolism in rusted leek leaves. *Physiol. Mol. Plant Pathol.* **32**, 229–235.

Rosenthal, G.A. (1983) A seed-eating beetle's adaptations to a poisonous seed. *Sci. Am.* Nov., 164–171.

Rosenthal, G.A. and Janzen, D.H. (1979) *Herbivores: Their Interaction with Secondary Plant Metabolites*. Academic Press, New York.

Ryan, C.A. (1990) Protease inhibitors in plants: Genes for improving defenses against insects and pathogens. *Ann. Rev. Phytopathol.* **28**, 425–449.

Schoeneweiss, D.F. (1975) Predisposition, stress and plant disease. *Ann. Rev. Phytopathol.* **13**, 193–211.

Seitz, E.W. and Hochter, R.M. (1964) Free proline in normal and crown gall tumor tissue of tomato and tobacco. *Life Sci.* **3**, 1033–1037.

Sequeira, L. (1983) Mechanisms of induced resistance in plants. *Ann. Rev. Microbiol.* **37**, 51–79.

Siddiqui, W.M., Halisky, P.M. and Lund, S. (1968) Relationship of clipping frequency to root and crown deterioration in red clover. *Phytopathol.* **58**, 486–488.

Siegel, M.R., Latch, G.C.M. and Johnson, M.C. (1987) Fungal endophytes of grasses. *Ann. Rev. Phytopathol.* **25**, 293–315.

Sinden, S.L., Schalk, J.M. and Stoner, A.K. (1978) Effect of daylength and maturity of tomato plants on tomatine content and resistance to the Colorado potato beetle. *J. Am. Soc. Hortic. Sci.* **103**, 596–600.

Singh, D. and Smalley, E.B. (1969) Changes in amino acid and sugar constituents of the xylem sap of American elm following inoculation with *Ceratocystis ulmi*. *Phytopathol.* **59**, 891–896.

Staley, J.M. (1965) Decline and mortality of red and scarlet oaks. *Forest Sci.* **11**, 2–17.

Steiner, W.E. (1984) A review of the biology of Phalacrid beetles (Coleoptera). In: *Fungus-Insect Relationships: Perspectives in Ecology and Evolution* (eds Q. Wheeler and M. Blackwell). Columbia Univ. Press, New York, pp. 424–445.

Stephens, G.R. (1971) The relation of insect defoliation to mortality in Connecticut forests. *Conn. Agric. Exp. Stn. Bull.* **723**, pp. 1–15.

Tallamy, D.T. and Krischik, V.A. (1989) Variation and function of cucurbitacins in *Cucurbita*: an examination of current hypotheses. *Am. Nat.* **133**, 766–786.

Thresh, J.M. (1981) The role of weeds and wild hosts in the epidemiology of plant virus diseases. In: *Pests, Pathogens and Vegetation* (ed. J.M. Thresh). Pitman Advanced Publishing Program, London, pp. 53–70.

Tu, J.C. and Ford, R.E. (1970) Free amino acids in soybeans infected with soybean mosaic virus, bean pod mottle virus, or both. *Phytopathol.* **60**, 660–664.

Tukey, H.B. (1970) The leaching of substances from plants. *Ann. Rev. Plant Physiol.* **25**, 305–324.

Vining, L.C. (1990) Functions of secondary metabolites. *Ann. Rev. Microbiol.* **44**, 395–427.

Walters, D.R. and Ayres, P.G. (1980) Effects of powdery mildew disease on the uptake and metabolism of nitrogen by roots of infected barley. *Physiol. Plant Pathol.* **17**, 369–379.

Wargo, P.M. and Houston, D.R. (1973) Infection of defoliated sugar maple trees by *Armillaria mellea*. *Phytopathol.* **63**, 209.

Wennström, A. and Ericson, L. (1991) Variation in disease incidence in grazed and ungrazed sites for the system *Pulsatilla pratensis – Puccinia pulsatillae*. *Oikos* **60**, 35–39.

White, T.C.R. (1974) The abundance of invertebrate herbivores in relation to the availability of nitrogen in stressed food plants. *Oecologia* **63**, 90–105.

Wicklow, D.T. (1988) Metabolites of fungal chemical defence systems. In: *Coevolution of Fungi with Plants and Animals* (eds K.A. Pirozynski and D.L. Hawksworth). Academic Press, London, pp. 173–201.

Wood, B.W., Tedders, W.L. and Reilly, C.C. (1988). Sooty mold fungus on pecan foliage suppress light penetration and net photosynthesis. *Hortic. Sci.* **23**, 851–853.

Yarwood, C.E. (1959) Predisposition. In: *Plant Pathology* (eds J.G. Horsfall and A.E. Dimond). Academic Press, New York, pp. 521–562.

Zimmerman, M.H. and McDonough, J. (1978) Dysfunction in the flow of food. In: *Plant Disease: an Advanced Treatise III* (eds J.G. Horsfall and E.B. Cowling). Academic Press, New York, pp. 117–140.

11

Modelling the effects of foliar pests and pathogens on light interception, photosynthesis, growth rate and yield of field crops

W.A.H. Rossing, M. van Oijen, W. van der Werf, L. Bastiaans and R. Rabbinge

11.1 Introduction

Differences in crop growth rate between years or locations may be explained by differences in the amount of light (solar radiation, 400–700 nm) received by the crop. However, variation in crop growth rate and yield may also be caused by other factors. According to resource levels and the presence or absence of injurious factors, three yield levels can be defined: the potential yield level, the attainable level and the actual level (Figure 11.1; Zadoks and Schein, 1979).

Potential yields are attained when crops grow with an ample supply of water and nutrients, while injurious biotic and abiotic agents are absent. Such situations are rare and may be obtained only in protected cultivation. Under such conditions, yields depend on site-specific abiotic conditions and crop physiological characteristics. Site factors include the courses of solar radiation, temperature and humidity over the year and over the day, the CO_2 concentration in the air and the physical soil characteristics. Major crop-dependent factors comprise phenology, three-dimensional structure of the canopy, temporal assimilate allocation pattern and the biochemical pathway of CO_2 fixation (C_3, C_4 or Crassulacean acid metabolism (CAM)). Together these factors constitute the growth- and yield-defining factors. Potential growth rates in a temperate, humid climate, e.g. in The Netherlands, appear to be in the order of 25 g dry

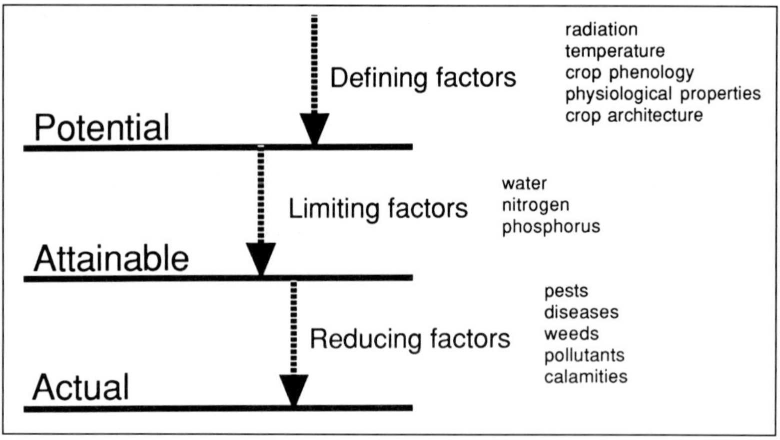

Figure 11.1. *Levels of crop production.*

matter (DM) m^{-2} day^{-1}. Expressed per unit of light, by definition the only limiting resource under optimal conditions, the growth rate is approximately 3 μg (DM) J^{-1} (light).

Shortage of water, nutrients, or both, limits yield to the attainable level. In addition to uptake of CO_2, which is used in photosynthesis, transpiration of water takes place through the stomata. The rate of transpiration depends on radiation, the drying power of the air (vapour pressure deficit) and stomatal aperture. The ratio of transpiration and CO_2 assimilation, the transpiration coefficient, is about 150–300 g (water) g^{-1} (DM) for crops in temperate climates. Thus, to maintain potential growth rates, 4000–8000 g (water) m^{-2}, or 4–8 mm, must be available for transpiration each day. Rates of CO_2 assimilation needed for potential crop growth rates can only be maintained at nitrogen concentrations of approximately 6% of the leaf dry matter (e.g. van Keulen and Seligman, 1987). The amount of nitrogen needed to support potential growth of a crop with a leaf area index of 4 m^2 (of leaf) m^{-2} (of ground) and a specific leaf area of 20 m^2 (of leaf) kg^{-1} (leaf DM) is 12 g m^{-2}. When less nitrogen is available in the leaf, the rate of photosynthesis is reduced. The figures presented are rules of thumb. Methods for estimating water- or nutrient-limited growth rates are described in more detail by van Keulen and Wolf (1986).

Pests, diseases, weeds, extreme weather conditions or pollutants reduce yield to the level which is actually realized in the field (Figure 11.1). The size of the yield reduction, i.e. damage (Zadoks, 1985), depends on the plant processes affected by a growth-reducing factor, i.e. injury components (after Zadoks, 1985), the growth rate of the healthy crop, and the timing and the intensity of growth reduction. The timing and the intensity of growth reduction due to biotic causes depends on the life cycle and population dynamics of the particular agent. Dynamic crop growth models offer a valuable framework within which to structure thinking about the interactions between growth-reducing factors and crop growth processes under variable environmental conditions.

Table 11.1. *Principal processes in plants which can be affected by growth-reducing factors*

Processes	Growth-reducing factors
Carbon and energy economy	
Photosynthesis	
Light interception and distribution	Weeds, necrotrophic fungi, leaf-eating insects
CO_2 diffusion CO_2 fixation other chloroplast processes	Fungal and viral pathogens, phloem-feeding insects and mites, SO_2
Respiration	
Maintenance	Many fungal and viral pathogens, mites, phloem-feeders
New syntheses (growth)	
Photorespiration	Leaf diseases
Allocation	Biotrophic fungi, phloem-feeders, SO_2, viruses
Transport	Phloem-feeding insects
Water economy	
Uptake	Root pathogens, nematodes
Transport	Vascular wilt diseases: *Verticillium*, *Fusarium*
Transpiration; stomatal regulation	Leaf diseases
Nutrient economy	
Uptake	Root pathogens, nematodes
Transport	Vascular wilt diseases: *Verticillium*, *Fusarium*
Redistribution	Leaf diseases
Morphogenesis	
Organ initiation	Mycoplasmas, galling aphids
Organ growth	Most growth-reducing agents (per def.)

Crop growth models can include process descriptions at many different levels of organization, e.g. molecular, cellular, tissue, organ, plant, crop or even agro-ecosystem. In studies of crop injury and damage, the levels of interest are those of the crop, the plant and its organs. When lower levels are included in models, problems arise due to lack of knowledge and large differences in the temporal and spatial scales of the processes (de Wit and Penning de Vries, 1982). At the same time, the explanatory power of the model may not increase substantially, or even diminish due to loss of model lucidity. For these reasons, many of the mechanisms of injury studied by physiological plant pathologists, e.g. differences in photosynthesis rate or turgor relations within areas of an infected leaf (reviews in Ayres, 1981; Pegg and Ayres, 1987; Scholes, Chapter 7), are often not used in crop growth models. In crop growth models, processes are more often defined at the plant level, a crop consisting of a population of plants. Here, four major functional subsystems can be distinguished; the economies of carbon, water and nutrients, and morphogenesis. Injury at the plant level can be classified according to the subsystem affected and the processes affected within each subsystem (Table 11.1). A phenomenological classification of injury components was given by Boote *et al.* (1983) who distinguished tissue consumers, leaf senescence accelerators, stand reducers, light stealers, photosynthetic rate reducers, assimilate sappers, and turgor reducers.

Explanatory crop growth models represent powerful tools to gain more insight into the consequences of different injury components for damage, thus allowing simple management rules to be derived. In recent years, two types of crop growth model have frequently been put forward for analysing the consequences of injury for crop growth and yield. These models differ mainly in the degree of detail to which light interception and crop photosynthesis are represented. These models are described here and applied to examples from two studies at Wageningen. The purpose of the comparison is to highlight the merits and limitations of each approach.

11.2 Two approaches to analysing damage with crop growth models

11.2.1 *Summary approach: light interception and utilization by the injured crop*

In many crops growing at the potential production level, growth rate is approximately linearly related to light intercepted by green foliage (Monteith, 1977). Thus, when total biomass is plotted against cumulative intercepted light a straight line is obtained. The slope of this line represents the average crop light-use efficiency (CLUE). The CLUE has been shown to be fairly constant over the growing period (Biscoe and Gallagher, 1977), provided that the average is calculated over periods of at least 14 days (Spitters, 1990). This relation between light interception by green area (LI) and biomass can be used for analytic and integrative purposes.

When analysing experimental results, differences between treatments can be attributed to LI or CLUE, thus distinguishing between effects on photosynthetic area and activity per unit green leaf area, respectively. When used for integrative purposes, LI and CLUE form the basis for a summary model of crop growth in which daily growth rate is calculated as the product of daily LI and CLUE. Total dry matter is obtained by integrating the daily growth rate over time in the growing season. Partitioning factors can be applied to account for the fractions of total biomass allocated to the various plant organs (Spitters, 1987; Russell *et al.*, 1989).

Damage by foliar pests and diseases may be due to reduction of LI, reduction of CLUE, or both (Figure 11.2). When CLUE is unaffected by injury, the photosynthetic capacity of the remaining green area is likely to be unaltered. Injury then simply causes a reduction of the amount of energy available for dry matter formation and damage is proportional to the amount of energy foregone. However, when CLUE is decreased, production capacity per unit of green area is affected and the relation between damage and cumulative intercepted light is more complex.

Examples of studies in which the LI/CLUE concept has been used to analyse injury and damage are given by Haverkort and Bicamumpaca (1986), Johnson (1987), Waggoner and Berger (1987) and van Oijen (1991b). Waggoner (1990) gives formulae to account for vertical and horizontal gradients of pests and pathogens in a crop when calculating LI.

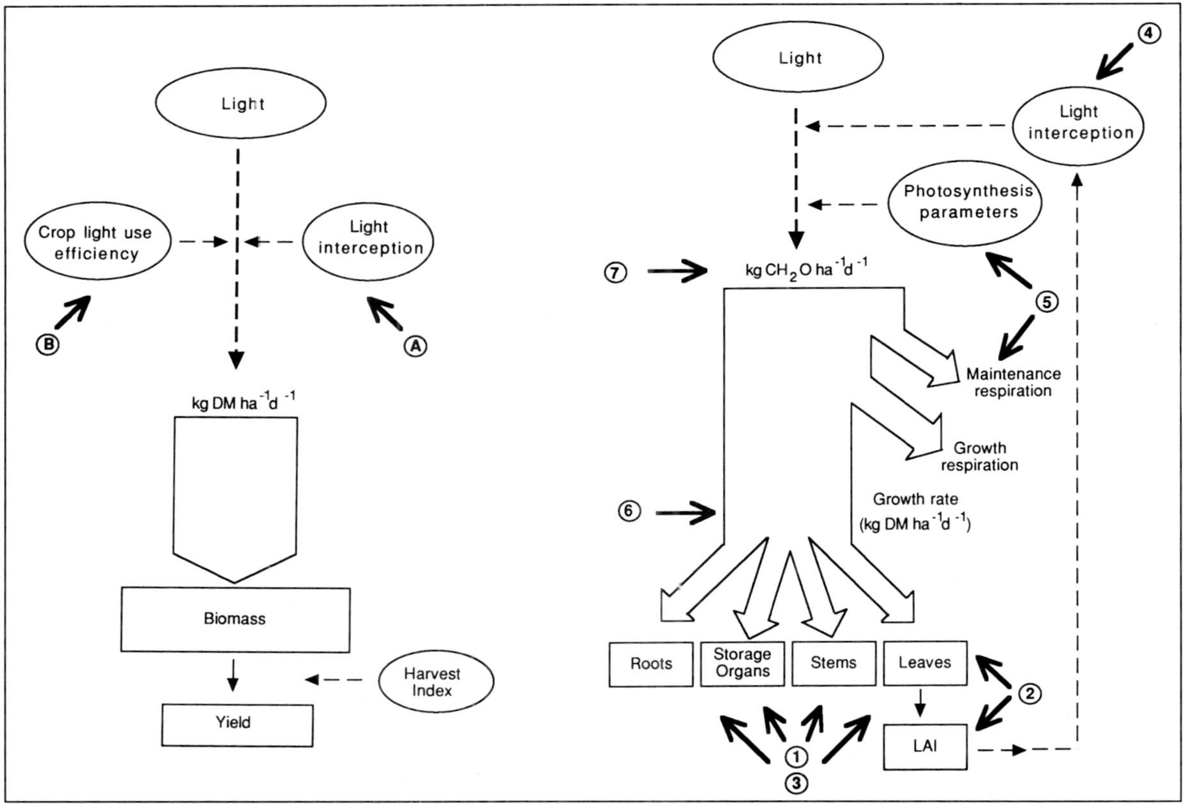

Figure 11.2. *Schematic representation of two explanatory approaches to modelling crop growth and injury (after Spitters, 1990). On the left the summary model of crop growth based on light interception, crop light-use efficiency and harvest index. On the right the comprehensive approach utilizing the light profile within the canopy, photosynthesis characteristics of individual leaves, respiration and dry-matter partitioning factors. Arrows indicate potential components of injury. In the summary model these are: effects on light interception (A) and utilization (B) by the crop. In the comprehensive model these are: tissue consumption (1), leaf senescence acceleration (2), stand reduction (3), light theft (4), (net) photosynthetic rate reduction (5), assimilate sapping (6), and turgor reduction (7).*

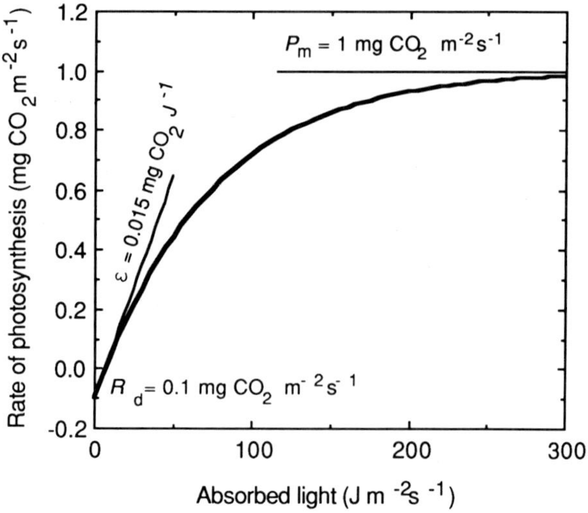

Figure 11.3*. Illustration of a photosynthesis* versus *light response curve, characterized by the parameters* P_m, ε *and* R_d, *see text. The curve is described by a negative exponential equation (Goudriaan, 1982).*

11.2.2 *Comprehensive approach: light interception and utilization by the injured plant*

A comprehensive approach to crop growth starts at the whole plant level. Light utilization by individual leaves, measured in detailed experiments, is combined with the light profile within the crop to arrive at estimates of daily crop growth rates. Light utilization is described by the light response curve, characterized by an initial slope, the initial light use efficiency of single leaves, ε, an asymptote at high light intensities, P_m, and an intercept representing the rate of dark respiration, R_d (Figure 11.3). Experimental analysis of injury components affecting light utilization is in terms of effects on the parameters describing the light response curve of individual leaves. The consequences of altered light utilization by leaves for crop growth are evaluated with a model. In the model (e.g. SUCROS, Spitters *et al.*, 1989) the vertical light profile is calculated using, basically, Beer's law for the penetration of three distinct light fluxes in the canopy: direct, diffuse and scattered light. At selected depths within the canopy and at selected times of the day, the rate of gross CO_2 assimilation is calculated using the response of individual leaves to light. The rate of gross CO_2 assimilation by the crop during 1 day is found by integration over the layers within the canopy and over time within the day. The rate of daily dry matter production is found by subtracting the rate of maintenance respiration from the calculated gross assimilation rate and accounting for the costs of allocation of the assimilates to various organs, and conversion into structural biomass. Integration over the days in the growing season results in total dry matter production.

The injury components of a growth-reducing factor can be introduced into SUCROS at the level of the whole plant (Figure 11.2). Vertical gradients of pests and pathogens can be taken into account by distinguishing different leaf layers and, within each layer, the categories 'healthy', 'injured' and 'dead' leaf area. Within the category 'injured', the intensity of injury, expressed for example as severity, i.e. the fraction of leaf area covered by lesions, is related to leaf photosynthesis. Recently, a universal approach to describe the effect of pests and pathogens on CO_2 assimilation by single leaves has been put forward (Bastiaans, 1990). The approach was developed for lesions of the leaf pathogen *Pyricularia oryzae* in rice, which decreases both the initial light use efficiency (ε) and the maximum rate of CO_2 assimilation (P_m) beyond the visual borders of a lesion. To relate the photosynthesis parameters to disease severity, 'virtual lesions' are introduced in which CO_2 assimilation is absent while in the remaining leaf area it is unaffected. This idea results in a standard function to describe the relation between disease severity and photosynthesis which is characterized by a single parameter, β, the ratio of virtual and visual leaf area. Values of β exceeding 1.0 imply that the rate of leaf photosynthesis is reduced more than can be explained by a reduction of green area and a loss of light intercepted by the visual lesion. Apparently, the 'zone of influence' of the pathogen extends beyond the borders of the visual lesion, possibly due to the presence of a mobile toxin. Biochemical studies complementary to photosynthesis measurements at the whole leaf level are needed to elucidate the causes of the existence of virtual lesions. The β-model can be used in pathosystems where injury causes clearly confined symptoms. van der Werf *et al.* (1990) compiled severity–photosynthesis relations from the literature and found for cereal leaf diseases β-values ranging from 1.26 ± 0.16 (brown rust) to 5.8 ± 0.6 (mildew), for *Cercospora* leafspot in peanut a β-value of 11.0 ± 3.5, and for brown mite on avocado a value of 1.3 ± 0.7.

Introducing the quantitative effects of the injury components into the crop growth model allows analysis of the consequences on crop growth and yield. Thus, damage is explained using the economies of carbon, water and nutrients, and plant morphogenesis as explanatory levels. Examples of analysis of damage using information on injury components at the whole plant level are given by Rabbinge and Rijsdijk (1981), Boote *et al.* (1983), Kropff (1988), van der Werf (1988), Kropff and Goudriaan (1989), Bastiaans (1990), Daamen and Jorritsma (1990), van Roermund and Spitters (1990) and Rossing (1991).

11.3 Analysis of injury and damage caused by foliar attack – three examples

11.3.1 *Introduction*

The two concepts of crop growth are now used to analyse experiments with three pathosystems. Potato late blight, *Phytophthora infestans*, may reduce yield directly when sporangia are rain-washed into the soil from sporulating leaf lesions and infect the tubers. However, yield loss is usually the result of the rapid destruction of the foliage, which results in shortening of the growing season.

Beet yellows virus (BYV) belongs to the closterovirus group and is transmitted by aphids, especially *Myzus persicae*, in a semi-persistent manner (Bar-Joseph *et al.*, 1979). Following transmission, the virus multiplies in the inoculated leaf. After some time, the virus is transported to the growing tissues, leaves and roots. Symptoms develop on the inoculated leaf and on the systemically infected leaves. Mature non-inoculated leaves do not become systemically infected (van der Werf *et al.*, 1989a), presumably because there is no, or negligible, phloem transport of virus to these assimilate-exporting leaves. Yellowing symptoms develop after an infected leaf is fully grown (van der Werf *et al.*, 1989b). Until these symptoms appear, the rate of photosynthesis is not markedly affected. Thus BYV-infected plants in the Dutch climate often have three distinct whorls of leaves: (1) an inner whorl of young leaves which are systemically infected but still green and photosynthetically active; (2) an intermediate whorl of mature systemically infected leaves which are yellow and photosynthetically impaired (Hall and Loomis, 1972a,b; Van der Werf, 1988); (3) an outer whorl of healthy mature and senescing leaves that appeared before the plant became infected and which are photosynthetically unimpaired.

Beet mild yellowing virus (BMYV) belongs to the luteovirus group and is persistently transmitted by *M. persicae.* The effects of this virus on sugarbeet have much in common with those of BYV. Leaves turn yellow, exhibit decreased rates of photosynthesis (van der Werf, unpublished results) and occur in whorls similar to those in BYV-infected plants. However, the symptoms take somewhat longer to develop (van der Werf, 1988).

11.3.2 *Application of the summary approach to crop growth*

The potato–late blight pathosystem: data and results. The effect of *P. infestans* on light interception and tuber yield was evaluated in a field experiment in 1988 with three cultivars, the early, susceptible cultivar Bintje, the early, resistant cultivar Surprise, and the late, resistant cultivar Pimpernel. Four different combinations of inoculum and contact fungicide were applied to obtain different levels of disease. In the inoculated treatment, a suspension of sporangia of *P. infestans* was applied about 2 months after planting and no fungicide was applied. In the control treatment no inoculum was applied and the fungicide maneb/fentin acetate was sprayed weekly, from leaf appearance until harvesting. In the treatment 'unsprayed-A', the weekly fungicide application was stopped at the same time as inoculum was applied in the inoculated treatment, while in the 'unsprayed-B' treatment, fungicide application was stopped 3 weeks later. Further details are given by van Oijen (1991b).

Infection led to yield losses in the inoculated, unsprayed-A and unsprayed-B treatments (Table 11.2). Percentage yield loss was always least in cv. Surprise and usually greatest in cv. Bintje.

Infection accelerated the onset of ground cover decline by up to 5 weeks, but had no effect on the rate of decline (Figure 11.4). Light interception accumulated over the growing season (LI) was calculated for each plot, assuming ground cover to be equivalent to light interception (Burstall and Harris, 1983). Relations between tuber yield and cumulative light interception are described by straight lines with a common slope (CLUE) of 2.06 g (tuber DM) MJ^{-1} (light) (Figure 11.5). No differences in slope were detected between cultivars or treatments.

Table 11.3 lists estimates of CLUE for potato crops made by different authors and from different experiments. In no experiment did *P. infestans* significantly affect CLUE. All experiments were conducted under optimum potential growth conditions for the

Table 11.2. *Final yield of tuber dry matter (t ha^{-1})*

Cultivar	Treatment			
	Inoculated	Unsprayed-A	Unsprayed-B	Control
Bintje	2.0	5.7	5.6	7.8
Surprise	4.2	6.4	7.4	7.7
Pimpernel	3.2	6.0	7.1	8.7

Least significant difference (LSD, $P = 0.05$) for pairwise comparison of cultivar means within treatment levels is 1.5 t ha^{-1}. Source: van Oijen (1991b).

healthy crop. Nevertheless, values for crop light-use efficiency varied between 1.8 and 3.2 g (tuber DM) MJ^{-1} (light).

Model calculations. Late blight reduces the cumulative light interception in two ways: first, by covering leaves with lesions, and, second, by accelerating leaf senescence in the remaining, green area. Because late blight occurs mainly on the lower leaves, damage due to shading of healthy leaf area by dead leaf area is unimportant. To evaluate the contribution of these two injury components to damage, the LI/CLUE crop growth model was parameterized for potato (Spitters and Schapendonk, 1990) and coupled to a model of *P. infestans* population dynamics (van Oijen, 1989). Acceleration of leaf senescence by the disease was included in the model. At any time during the epidemic, disease severity, expressed as the percentage of leaf area covered by lesions, induced an equal percentage of leaf senescence in the remaining, lesion-free leaf area (van Oijen, 1991a). In addition, leaf area was lost as a result of normal, physiological ageing. The performance of the model was tested by simulating the course of ground cover in the four treatments, using measured values of average daily

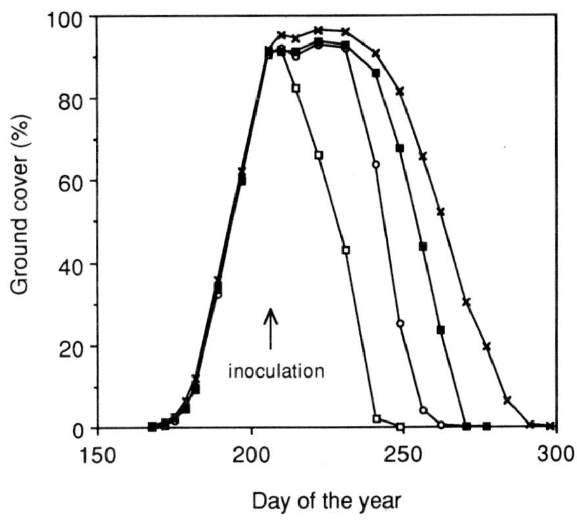

Figure 11.4. *Time course of ground cover of potato. Points represent means over cultivars within treatments. Inoculated (——□——), unsprayed-A (——○——), unsprayed-B (——■——) and control (——×——). Reproduced from van Oijen (1991b) with permission from the European Association for Potato Research.*

Figure 11.5. Relation between yield of tuber dry matter and cumulative light interception by green area of three cultivars. For each cultivar results of four treatments at three harvest dates are shown. Points represent averages in four replicates, regression lines are based on individual measurements. Cultivars are Bintje (○), Surprise (●) and Pimpernel (×). Reproduced from van Oijen (1991b) with permission from the European Association for Potato Research.

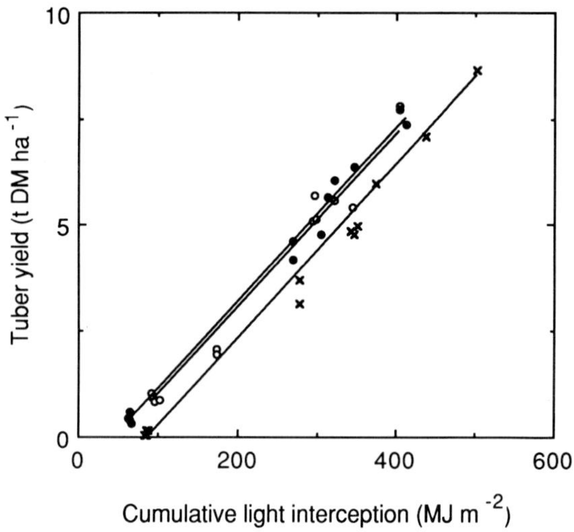

temperature and light, and a CLUE of 2.35 g MJ^{-1} for dry matter production as measured in the experiment. The agreement between simulations and measurements was excellent for most cultivars and treatments (Table 11.4). The only exception was the overestimation of green leaf area duration of the late cv. Pimpernel in the absence of disease. Yield was estimated well in all cases.

The contribution of disease-induced leaf senescence to yield loss was assessed by re-running the model with normal, physiological ageing as the only cause of leaf senescence. The results showed that disease-induced leaf senescence accounted for 4–15% of yield loss by *P. infestans*. Thus, the major fraction of yield loss was caused by loss of green leaf area due to lesion extension.

Discussion. Light-use efficiency of the potato crop, averaged over the growing season, was similar for different cultivars and was not affected by the different levels of infection by *P. infestans*. Thus, differences in yield between treatments were caused mainly by differences in the amount of light intercepted.

In experiments similar to the one described here, but at different locations and with

Table 11.3. *Estimates of light use efficiency of potato crops in different experiments*

Crop light use efficiency (g (tuber DM) MJ^{-1} (light))	Authors	Location
3.17	van Oijen, 1991b	The Netherlands
2.06	van Oijen, 1991b	The Netherlands
1.81	van Oijen, 1991b	The Netherlands
3.06	Haverkort and Bicamumpaka, 1986	Rwanda
~2.60	Haverkort and Bicamumpaka, 1986	Rwanda
3.20	Waggoner and Berger, 1987	Israel

Table 11.4. *Comparison of cumulative light interception (LI) and yield of tuber dry matter in field experiments and simulations*

Cultivar	Treatment	LI (MJ m^{-2})		Yield (t ha^{-1})	
		Measured	Simulated	Measured	Simulated
Bintje	Control	403	448	7.8	7.9
	Inoculated	174	212	2.0	2.4
Surprise	Control	404	448	7.7	7.9
	Inoculated	270	298	4.2	4.4
Pimpernel	Control	502	554	8.7	9.0
	Inoculated	279	299	3.2	3.0
r^2		0.99		0.99	
$\sqrt{\text{MSE}}$		39		0.3	

Two measures of the degree of agreement between simulated and measured values are shown: the coefficient of determination, r^2, and the square root of the mean squared error, $\sqrt{\text{MSE}}$. Source: van Oijen (1991b).

20 potato cultivars, van Oijen (1990) never found a significant change in light-use efficiency due to late blight. These results corroborate the findings of Haverkort and Bicamumpaka (1986) in experiments with tropical highland potato crops and late blight in Rwanda. Similarly, Waggoner and Berger (1987), re-analysing data of Rotem *et al.* (1983), concluded that CLUE of potato was unaffected by disease.

The estimates of CLUE obtained in different experiments show a large variation. In all experiments, crops were well supplied with water and nutrients. However, variation in growth rates may have been caused by factors such as late planting time and destruction of leaves by frost (van Oijen, 1991b), or water shortage at mid-day (Haverkort, personal communication, 1992).

Assessment of the relative importance of the injury components causing reduced light interception in infected crops was possible only by using the model. Knowing that lesion growth is the major injury component enabled van Oijen (1991a) to suggest that breeding strategies for late blight-tolerant potato varieties might benefit from putting greater emphasis on lesion growth rate as a selection criterion.

The BYV and BMYV pathosystems: data and results. In 1989 field plots of sugarbeet were either infected with BYV or BMYV, or treated regularly with an aphicide to produce uninfected controls. Infection was carried out by transferring 5–10 infectious *M. persicae* onto each plant at the cotyledon stage. At 2-week intervals from emergence to harvest, ground cover by green leaf area was estimated using a grid. As with potato, the leaf orientation of young sugarbeet plants tends to be planophile, such that the proportion of ground covered is approximately equivalent to the fraction of incident light intercepted (van der Werf, unpublished). In older plants the inner leaves are more erect.

Total biomass, root biomass and sugar yield were reduced by infection with BYV or BMYV (Table 11.5), BYV causing the greater effects. In spite of the early time of

Table 11.5. *Total dry matter, root dry matter and sugar yield in the 1989 experiment*

Treatment	Yield ($t\,ha^{-1}$)		
	Total biomass	Root biomass	Sugar yield
Control	16.0	7.8	5.1
BMYV-infected	14.0	6.0	4.0
BYV-infected	9.0	3.3	2.1
LSD[1]	1.6	0.7	0.4

[1] Least significant difference (LSD, $P=0.05$) for pairwise comparison of yield means between treatment levels.

infection, the first symptoms did not appear until after the exponential growth phase (Figure 11.6). This was due to the long incubation period of the viruses (approximately 3 weeks for BYV and 5 weeks for BMYV at early crop development stages) (van der Werf *et al.*, 1989b). As infection was at the cotyledon stage, there were no healthy uninfected leaves on these plants (see Section 11.3.1). The green leaf area of plants consisted entirely of young systemically infected leaves which had not yet developed symptoms. The fraction of yellow leaves increased during the growing season due to the decreasing rates of leaf appearance and expansion, and due to symptom development on the older, fully grown leaves. Ground cover by yellow leaves after BMYV infection was lower than after BYV infection at all times.

CLUE of the BMYV-infected treatment (1.68 g (DM) MJ^{-1} ± 0.07 (s.e.)) did not differ significantly from the control (1.62 g (DM) MJ^{-1} ± 0.04 (s.e.)). Infection with BYV significantly reduced the slope of the relations (CLUE was 1.46 g (DM) MJ^{-1} ± 0.04 (s.e.)).

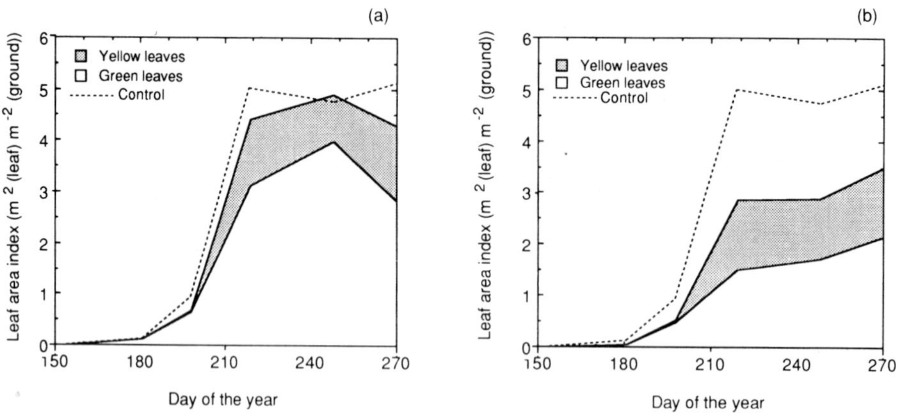

Figure 11.6. *Time course of ground cover of sugarbeet infected with BMYV (a) or BYV (b), and controls in the 1989 field experiment.*

Discussion. Although both BYV and BMYV reduced total biomass, root biomass and sugar yield, the nature of their interference with crop growth processes appears to be different. BMYV did not affect the light use efficiency of the crop. Therefore, damage due to BMYV appears attributable only to a reduction in the amount of light intercepted by green leaves. BYV, in contrast, reduced both the amount of intercepted light and the efficiency of light use.

11.3.3 *Application of the comprehensive approach to crop growth*

The potato–late blight pathosystem: data and results. Cultivars Bintje and Surprise were grown outdoors in pots, inoculated in a greenhouse 5 weeks after planting, and placed in a climate chamber after a further 2 weeks. CO_2 exchange was measured on relatively young, uninoculated leaves at different irradiances (details are given by van Oijen, 1990). Disease severity, expressed as the percentage of leaf or stem area below the measured leaf covered by lesions, varied between 10 and 20% in the inoculated plants and was 0% in the controls. In all inoculated plants at least one lesion completely encircled the stem at some point below the measured leaf. Nevertheless, infection did not significantly affect P_m, ε or R_d of either cv. Bintje or cv. Surprise.

Discussion. The photosynthetic capacity of green leaves was not affected by *P. infestans*. Thus, the virtual lesion area of *P. infestans* appears to be equal to the visual lesion area. These findings support the results of the analysis with the summary model: *P. infestans* reduces light interception, but does not affect light utilization within the green leaf area. Further analysis with the comprehensive crop growth model is unnecessary.

The sugarbeet–BYV and sugarbeet–BMYV pathosystems: data and results. van der Werf (1988) measured the effect of BYV on CO_2 assimilation and respiration in the field. Leaves of infected plants were classified as 'uninfected' (whorl 3 as described in Section 11.3.1), 'infected but without symptoms yet' (whorl 1 as in Section 11.3.1), 'with cleared veins', 'greenish yellow', or 'bright yellow'. Using nylon gauze to create a range of shading allowed assessment of the photosynthesis–light response curve of individual leaves. The reflection and transmission of light by healthy and infected leaves was measured on one occasion (details are given by van der Werf, 1988).

Infection with BYV reduced P_m and ε, and increased R_d in leaves showing yellowing symptoms, but had no effect on photosynthesis in uninfected leaves on the same plants (Table 11.6). Later measurements showed similarly that photosynthesis was not impaired in infected, green heart leaves (the 1988 data were inconclusive; van der Werf, unpublished results). In leaves which showed symptoms on only one half of the blade, the rate of photosynthesis was unaffected in the other, green half.

Absorption of light in uninfected, green leaves (88%) was similar to absorption in

Table 11.6. *Photosynthesis parameters (±s.e.) of uninfected sugarbeet leaves and different categories of BYV-infected leaves on infected plants*

	P_m (mg (CO_2) m^{-2} s^{-1})	ε (μg (CO_2) J^{-1} (light))	R_d (mg (CO_2) m^{-2} s^{-1})	*n*
Healthy plant	1.09 (0.04)	10.9 (0.3)	0.06 (0.01)	43
Infected				
Uninfected	1.03 (0.08)	11.0 (1.0)	0.03 (0.01)	8
Vein clearing	0.62 (0.05)	9.9 (1.2)	0.11 (0.03)	5
Greenish yellow	0.32 (0.04)	8.8 (1.0)	0.15 (0.02)	8
Bright yellow	0.16 (0.04)	7.8 (1.3)	0.11 (0.02)	13

Source: van der Werf (1988).

infected, green leaves (85%), but greater than in leaves showing yellowing symptoms (63–78%) (van der Werf, 1988).

Model calculations. The comprehensive crop growth model was parameterized to simulate the growth of sugarbeet in the 1989 field experiment with a standard parameter set (Spitters *et al.*, 1989). Four injury components were quantified to calculate the effect of virus infection on the plant: (1) reduced leaf expansion; (2) increased scattering of incident light by yellow leaves; (3) reduced photosynthesis rate in yellow leaves at high and low light intensities; and (4) increased respiration in yellow leaves. Although observed in BYV-infected plants, these injury components were assumed also to apply to the growth of plants infected with BMYV.

The first injury component was built into the model by introducing measured leaf area indices as a forcing function (Figure 11.7). The second injury component was introduced by calculating a weighted average scattering coefficient for the whole leaf canopy on the basis of the observed proportions of green and yellow leaf area (Figure

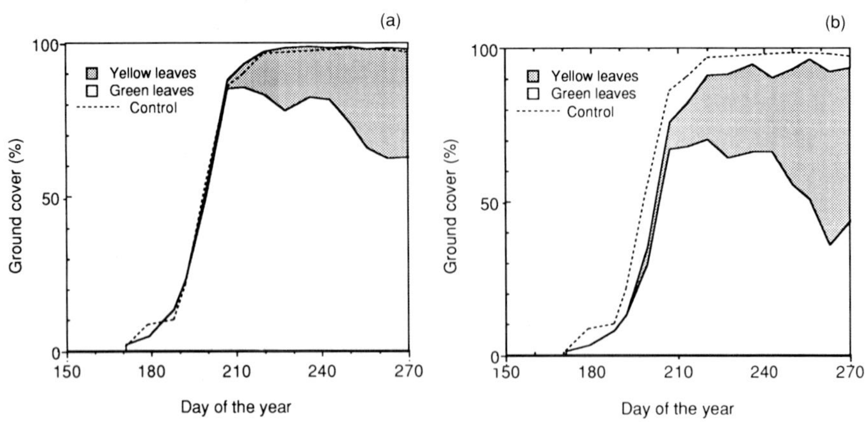

Figure 11.7. *Time course of leaf area index of sugarbeet infected with BMYV (a) and BYV (b), and control treatment in the 1989 field experiment.*

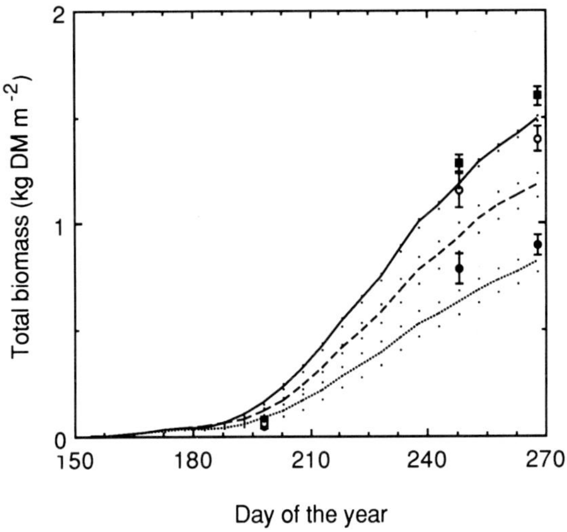

Figure 11.8. *Observed sugarbeet production in 1989 and simulation results obtained with the comprehensive crop growth model. The standard error of the simulation results (indicated by dots) was obtained by running the model with leaf area index data from individual plots. Field data: control (■), BMYV-infected (○) and BYV-infected (●). Simulation results: control (——), BMYV-infected (– – –) and BYV-infected (· · · ·).*

11.7) and measured scattering coefficients of 0.12 and 0.40, respectively (van der Werf, 1988). In the model only two types of leaves are distinguished, green and yellow. For the green leaves, the leaf photosynthesis parameters of Spitters *et al.* (1989) were used: P_m = 1.25 mg CO_2 m^{-2} s^{-1}; ε = 12.5 μg CO_2 J^{-1}. For yellow leaves (injury component 3), the parameter values were: P_m = 0.28 mg CO_2 m^{-2} s^{-1}; ε = 9.7 μg CO_2 J^{-1}. Increased respiration (injury component 4) was taken into account by assuming that yellow leaves exhibit a rate of maintenance respiration 2.5 times higher than that of normal healthy leaves. Details are given in van der Werf (1988).

The model slightly underestimated the real growth figures (Figure 11.8). Differences in yield between the three treatments were, however, described fairly well. Earlier sensitivity analysis of the model (Rabbinge *et al.*, 1990; Table 11.7) had shown that the reduction of photosynthetic capability of the yellow leaves was the principal injury component, explaining 70% of the yield reduction.

Discussion. In view of the size of the reduction of net CO_2 assimilation of individual leaves, the assumption that yellow leaves do not contribute to growth would appear a reasonable simplification. Thus, the detailed analysis suggests the simpler crop growth model is valuable for summarizing injury. Virus infection is expected mainly to affect cumulative light interception, since net CO_2 assimilation, and therefore light utilization, is not impaired in the green (parts of) leaves. Infection is expected to increase slightly the light-use efficiency of the infected crop as yellow leaves contribute somewhat to crop production.

Table 11.7. *Simulated relative contribution of components of injury by BYV to sugarbeet in a field experiment in 1986*

Injury components	Early infection		Late infection	
	Yield (%)	%damage[1]	Yield (%)	%damage
No disease	100		100	
		7.1		1.5
1 (reduced leaf area index)	92.9		98.5	
		5.2		0.2
1 + 2 (reduced light absorption)	88.0		98.2	
		36.4		1.7
1 + 2 + 3 (reduced photosynthesis, ε and P_m)	56.0		96.6	
		11.7		0.5
1 + 2 + 3 + 4 (increased respiration)	49.4		96.1	
Measured	48.2 ± 2.5		93.4 ± 5.1	

Early infection: 5 June 1986, seven leaves, LAI = 0.1. Late infection: 17 July 1986, 21 leaves, LAI = 5.1. More details are in Rabbinge *et al.* (1990).

[1] %damage is the decrease in yield (%) due to incorporating the next injury component in the model, relative to the previously calculated yield. The % damage due to injury component $x + 1$ is calculated as

$$100 - 100 \cdot \frac{\text{yield with injury components 1 through } x + 1}{\text{yield with injury components 1 through } x}.$$

The results of application of the summary model to the 1989 field experiment (Section 11.3.2) showed that conclusions from the detailed analysis held good only for BMYV. For BYV-infected crops, in contrast, the decrease in CLUE found in the field experiment could not be explained on the basis of the injury mechanisms in the detailed analysis. Stated simply, it could be said that green, BYV-infected leaves contribute less to crop production than green, healthy leaves. A plausible explanation would be that infection with BYV increases respiration, not only in the yellow leaves, but also in other plant parts such as the petioles, crown and storage root. Such an increase in whole plant respiration was not taken into account in the comprehensive model because reliable data concerning increased respiration in tissues other than yellow leaves were not available. Further analysis of injury components of BYV is therefore necessary.

11.4 Evaluation

Crop growth models are potentially useful tools for estimating the effects of pests and pathogens on crop growth and for studying the interactions of underlying mechanisms of damage with each other and with the growing conditions. This chapter has discussed two related but still distinct approaches to modelling crop growth by way of example. The first, summary, approach is based on cumulative LI by the visibly uninjured portion of the leaf canopy of the crop. In this approach, yield is calculated as the product of LI and CLUE. Several pests and pathogens affect yield by lowering

LI, while leaving CLUE unaffected, suggesting that the uninjured portion of the plant functions normally. *P. infestans* and BMYV are examples of such pathogens. Other pests and pathogens not only decrease LI, but also affect CLUE, indicating that the apparently uninjured portion of the plant does not function normally. Beet yellows virus is an example of such a pathogen. The causes of the effect of a pest or pathogen on the CLUE can be analysed with a comprehensive crop growth model like SUCROS. Measurements and simulations, using this second approach, suggest that increased respiration in BYV-infected plants may explain the lowered CLUE.

Although in the examples attention has been focused on pathogens, both approaches are equally suitable for analysing the effects of other biotic and abiotic growth reducing factors, as shown in studies of potato–leafhopper (Johnson *et al.*, 1987), winter wheat–grain aphid (Rossing, 1991), sugarbeet–weeds (Kropff, 1988) and faba bean–SO_2 (Kropff and Goudriaan, 1989). Pathosystem-specific aspects comprise the major injury components, and the temporal and spatial dynamics of attack.

The LI/CLUE approach to analysing injury and damage is appealing because of its conceptual simplicity and the ease of empirical parameter estimation. However, the value of CLUE, even under conditions allowing potential production, may vary considerably (Table 11.3, Russell *et al.*, 1989). Seasonal temperature and radiation patterns, and management practices affect CLUE as they affect photosynthesis or canopy structure. The variability of CLUE precludes application of the LI/CLUE model for predictive purposes outside the agro-ecological environment in which the estimates were derived. Fortunately, the variability of CLUE does not hamper the usefulness of the LI/CLUE model as a diagnostic and summarizing tool in studies of injury and damage. The LI/CLUE model allows two broad categories of injury components to be distinguished: injury components affecting LI and injury components affecting CLUE. For more detailed analysis, especially of the latter category, a more comprehensive approach to light utilization is required.

The comprehensive SUCROS-type approach to analysing injury and damage is conceptually more involved and requires sophisticated equipment for estimating the parameters of the photosynthesis–light response curve. Consequences of effects measured at the plant level are evaluated at the crop level by applying the model. Such evaluation must take into account that the model outcome is uncertain as a result of uncertainty in the parameters of the model. Methods for objective, statistical estimation of parameters have been developed (e.g. Klepper and Rouse, 1991), so that such comprehensive crop growth models can also be 'tuned' to specific environmental conditions for the purpose of developing tools for prediction. As a tool for structuring thought, and assessing the relative importance of different injury components, the comprehensive crop model is very suitable, because of its 'bottom-up' approach, i.e. it starts at the whole plant level. The uncertainty in the model outcome renders the comprehensive approach less appropriate for starting the process of identifying injury components, as was shown in the beet–BYV study.

The approach chosen in a specific study will depend on the questions to be answered. The LI/CLUE approach seems a good starting point for analysis. However, studies using a comprehensive approach will remain necessary to draw conclusions about the components of injury operating at the whole plant or lower levels of organization, as was shown in the beet–BMYV example. Studies at the plant level are also necessary to provide reasons for the variability of CLUE between studies, observed for instance in the work with potato crops.

References

Ayres, P.G. (ed.) (1981) *Effects of Disease on the Physiology of the Growing Plant.* Cambridge University Press, Cambridge.

Bar-Joseph, M., Garnsey, S.M. and Gonsalves, D. (1979) The closteroviruses: a distinct group of elongated plant viruses. *Adv. Virus Res.* **25**, 93–168.

Bastiaans, L. (1990) The ratio between virtual and visual lesion size as a measure to describe reduction in leaf photosynthesis of rice due to leaf blast. *Phytopathol.* **81**, 611–615.

Biscoe, P.V. and Gallagher, J.N. (1977) Weather, dry matter production and yield. In: *Environmental Effects on Crop Physiology* (eds J.J. Landsberg and C.V. Cutting). Academic Press, London, pp. 75–100.

Boote, K.J., Jones, J.W., Mishoe, J.W. and Berger, R.D. (1983) Coupling pests to crop growth simulators to predict yield reductions. *Phytopathol.* **73**, 1581–1587.

Burstall, L. and Harris, P.M. (1983) The estimation of percentage light interception from leaf area index and percentage ground cover in potatoes. *J. Agric. Sci.* **100**, 241–244.

Daamen, R.A. and Jorritsma, I.T.M. (1990) Effects of powdery mildew and weather on winter wheat yield. 2. Effects of mildew epidemics. *Neth. J. Plant Pathol.* **96**, 35–46.

Goudriaan, J. (1982) Potential production processes. In: *Simulation of Plant Growth and Crop Production* (eds F.W.T. Penning de Vries and H.H. van Laar). Pudoc, Wageningen, pp. 98–113.

Hall, A.E. and Loomis, R.S. (1972a) Photosynthesis and respiration by healthy and beet yellows virus-infected sugar beets (*Beta vulgaris* L.). *Crop Sci.* **12**, 566–572.

Hall, A.E. and Loomis, R.S. (1972b) An explanation for the difference in photosynthetic capabilities of healthy and beet yellows virus-infected sugar beets (*Beta vulgaris* L.). *Plant Physiol.* **50**, 576–580.

Haverkort, A.J. and Bicamumpaka, M. (1986) Correlation between intercepted radiation and yield of potato crops infested by *Phytophthora infestans* in central Africa. *Neth. J. Plant Pathol.* **92**, 239–247.

Johnson, K.B. (1987) Defoliation, disease, and growth: a reply. *Phytopathol.* **77**, 1495–1497.

Johnson, K.B., Teng, P.S. and Radcliffe, E.B. (1987) Analysis of potato foliage losses caused by interacting infestations of early blight, Verticillium wilt, and potato leafhopper; and the relationship to yield. *Z. Pflanzenkr. Pflanzenschutz* **94**, 22–33.

van Keulen, H. and Seligman, N.G. (1987) *Simulation of Water Use. Nitrogen Nutrition and Growth of a Spring Wheat Crop.* Pudoc, Wageningen.

van Keulen, H. and Wolf, J. (eds) (1986) *Modelling of Agricultural Production: Weather, Soils and Crops.* Pudoc, Wageningen.

Klepper, O. and Rouse, D.I. (1991) A procedure to reduce parameter uncertainty for complex models by comparison with real system output illustrated on a potato growth model. *Agric. Syst.* **36**, 375–395.

Kropff, M.J. (1988) Modelling the effects of weeds on crop production. *Weed Res.* **28**, 465–471.

Kropff, M.J. and Goudriaan, J. (1989) Modelling short-term effects of sulphur dioxide. 3. Effects of SO_2 on photosynthesis of leaf canopies. *Neth. J. Plant Pathol.* **95**, 265–280.

Monteith, J.L. (1977) Climate and the efficiency of crop production in Britain. *Phil. Trans. R. Soc. London, Ser. B* **281**, 277–294.

van Oijen, M. (1989) On the use of mathematical models from human epidemiology in breeding for resistance to polycyclic fungal leaf diseases of crops. In: *Parental Line Breeding and Selection in Potato Breeding* (eds K.M. Louwes, H.A.J.M. Toussaint and L.M.W. Dellaert). Pudoc, Wageningen, pp. 26–37.

van Oijen, M. (1990) Photosynthesis is not impaired in healthy tissue of blighted potato plants. *Neth. J. Plant Pathol.* **96**, 55–63.

van Oijen, M. (1991a) *Identification of the major characteristics of potato cultivars which affect yield loss caused by late blight.* Ph.D. dissertation. Wageningen Agricultural University.

van Oijen, M. (1991b) Light use efficiencies of potato cultivars with late blight (*Phytophthora infestans*). *Potato Res.* **34**, 123–132.

Pegg, G.F. and Ayres, P.G. (eds) (1987) *Fungal Infection of Plants.* Cambridge University Press, Cambridge.

Rabbinge, R. and Rijsdijk, F.H. (1981) Disease and crop physiology: a modeller's point of view. In: *Effects of Disease on the Physiology of the Growing Plant* (ed. P.G. Ayres). Cambridge University Press, Cambridge, pp. 201–220.

Rabbinge, R., Rossing, W.A.H. and van der Werf, W. (1990) The bridge function of production ecology in pest and disease management. In: *Theoretical Production Ecology: Reflections and Prospects* (eds R. Rabbinge, J. Goudriaan, H. van Keulen, F.W.T. Penning de Vries and H.H. van Laar). Pudoc, Wageningen, pp. 181–195.

van Roermund, H.J.W. and Spitters, C.J.T. (1990) Simulation of yield reduction by leaf rust in winter wheat, applied to the analysis of genetic variation in partial resistance. *Neth. J. Plant Pathol.* **96**, 17–28.

Rotem, J., Bashi, E. and Kranz, J. (1983) Studies of crop loss in potato blight caused by *Phytophthora infestans. Plant Pathol.* **32**, 117–122.

Rossing, W.A.H. (1991) Simulation of damage in winter wheat caused by the grain aphid *Sitobion avenae*. 2. Construction and evaluation of a simulation model. *Neth. J. Plant Pathol.* **97**, 25–54.

Russell, G., Marshall, B. and Jarvis, P.G. (1989) Absorption of radiation by canopies and stand growth. In: *Plant Canopies, their Growth, Form and Function* (eds G. Russell, B. Marshall and P.G. Jarvis). Cambridge University Press, Cambridge, pp. 21–39.

Spitters, C.J.T (1987) An analysis of variation in yield among potato cultivars in terms of light absorption, light utilization and dry matter partitioning. *Acta Hortic.* **214**, 71–84.

Spitters, C.J.T (1990) Crop growth models: their usefulness and limitations. *Acta Hortic.* **267**, 349–368.

Spitters, C.T.J. and Schapendonk, A.H.C.M. (1990) Evaluation of breeding strategies for drought tolerance in potato by means of crop growth simulation. *Plant Soil* **123**, 193–203.

Spitters, C.J.T, van Keulen, H. and van Kraalingen, D.W.G. (1989) A simple and universal crop growth simulator: SUCROS 87. In: *Simulation and Systems Management in Crop Protection* (eds R. Rabbinge, S.A. Ward and H.H. van Laar). Pudoc, Wageningen, pp. 147–181.

Waggoner, P.E. (1990) Defoliation, disease and growth. In: *Theoretical Production Ecology: Reflections and Prospects* (eds R. Rabbinge, J. Goudriaan, H. van Keulen, F.W.T. Penning de Vries and H.H. van Laar). Pudoc, Wageningen, pp. 149–180.

Waggoner, P.E. and Berger, R.D. (1987) Defoliation, disease and growth. *Phytopathol.* **77**, 393–398.

van der Werf, W. (1988) *Yellowing viruses in sugarbeet; epidemiology and damage.* Ph.D. dissertation, Wageningen Agricultural University.

van der Werf, W., Bastiaans, L., Rossing, W.A.H. and Rabbinge, R. (1990) Modeling crop response to growth reducing factors. In: *Proceedings of the International Symposium on Biotic Stress of Barley in Arid and Semi-Arid Environments.* Montana State University and ICARDA, pp. 87–102.

van der Werf, W., Bonnier, F. and Peters, D. (1989a) Retrospective estimation of the date of infection with beet yellowing viruses in sugar-beet under field conditions. *Neth. J. Plant Pathol.* **95**, 327–341.

van der Werf, W., Kempenaar, C. and Peters, D. (1989b) The incubation period of beet yellowing viruses in sugar-beet under field conditions. *Neth. J. Plant Pathol.* **95**, 241–258.

de Wit, C.T. and Penning de Vries, F.W.T. (1982) L'analyse des systèmes de production primaire. In: *La Productivité des Pâturages Sahéliens* (eds F.W.T. Penning de Vries and M.A. Djiteye). Agricultural Research Reports no. 918. Pudoc, Wageningen, pp. 20–27.

Zadoks, J.C. (1985) On the conceptual basis of crop loss assessment: the threshold theory. *Ann. Rev. Phytopathol.* **23**, 455–473.

Zadoks, J.C. and Schein, R.D. (1979) *Epidemiology and Plant Disease Management.* Oxford University Press, New York.

12

Constraints on the evolution of resistance to pests and pathogens

M.A. Parker

12.1 Introduction

Understanding the evolutionary constraints on resistance to parasitism is a complex problem, because the phenotypic expression of this trait depends on the interaction between two independent genomes: that of the host and that of the parasite. Since the mean level of resistance can change due to evolution in either species, this trait may evolve in a fashion that is not necessarily favourable for the host. To date, there is no consensus about what constraints are most important in preventing an evolutionary improvement of parasite resistance in natural plant populations. Indeed, there has been minimal discussion of what research strategies can best resolve this problem.

In this chapter, the term 'parasite' will be used to refer collectively to both metazoan herbivores and microbial pathogens that exploit plants. Two alternative research traditions on parasite resistance in plants will be reviewed that make fundamentally different assumptions about the nature of evolutionary constraints. The chapter begins by providing an operational definition of constraint, and then outlines the assumptions, goals, methods and biases of the two research strategies. Finally, it is argued that a rigorous understanding of constraints must involve a synthesis of basic elements from both of these research traditions.

12.2 Identifying constraints

Given that the optimal state for a plant is to be completely free from parasite attack, one might say that a constraint on resistance exists whenever a plant population is currently being exploited by at least one parasite taxon. However, this definition is too broad to be useful. No plant is absolutely free from parasitism; what we really want to understand is why some populations are relatively more vulnerable than others.

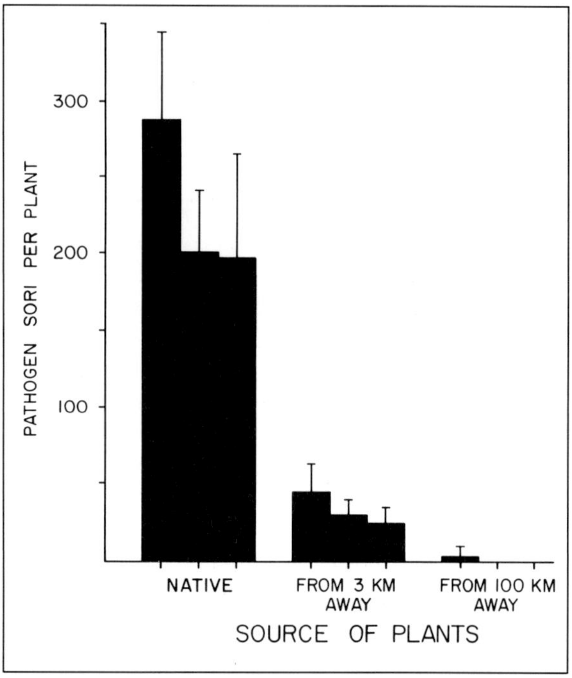

Figure 12.1. *Disease-free seedlings (germinated in a greenhouse) from each of three populations of* A. bracteata *were transplanted into randomly chosen locations in the native habitat of one of the populations. For each population, selfed progeny from three plants were used* (n = *10 to 16 transplants per progeny group). Data are the mean (+ 1 s.e.) of the number of* S. decipiens *sori (spore-producing structures) developing per plant in each progeny group after 53 days' growth. Native plants had significantly higher infection levels than plants from either of the other two populations (Kruskal–Wallis test, $P < 0.001$).*

Thus, for the purposes of this chapter, a constraint on resistance will be defined as any case where the mean phenotype in a plant population is shown to be less resistant than some alternative phenotype of the same species that currently exists or that can be experimentally constructed. The following two examples illustrate this concept. First, many plants have an inducible chitinase enzyme that serves to defend them from those fungal pathogens with chitin in their cell walls. The environmental cue that induces the accumulation of this enzyme is pathogen invasion. Broglie *et al.* (1991) created transgenic plants with a chitinase linked to a promoter causing constitutive enzyme expression, and found that these plants had significantly enhanced resistance to the fungal pathogen, *Rhizoctonia solani*. This implies that the resistance of the original plants could be improved by a simple change from inducible to constitutive expression of chitinase.

A second strategy for analysing constraints is to examine geographic variation within a species. Figure 12.1 shows the results of a field transplant experiment where plants from three isolated populations of the legume, *Amphicarpaea bracteata*, were simultaneously exposed to a local population of the pathogen, *Synchytrium decipiens* (Chytridiomycota). Plant genotypes native to this site were significantly more susceptible to infection than plants derived from the two foreign populations. Thus, for native plants, it is clear that an improvement in disease resistance is possible, because other plant genotypes of the same species already possess superior resistance to the pathogens found at this site.

Whether by genetic manipulation or by analysis of geographic variation, these examples show that we cannot automatically assume that all plants have acquired the maximum possible level of resistance to their parasites. In such cases, the central problem is to understand why evolution has not resulted in plants that are better defended.

12.3. Two research traditions

Historically, the most common approach to analysing constraints has been to assume that selective trade-offs exist between parasite resistance and other essential plant functions. This approach advocates that resistance can best be understood by a cost-benefit analysis, where the benefits of limiting parasite attack are balanced by various types of fitness costs associated with resistance (Whittaker and Feeny, 1971; Cates, 1975; Chew and Rodman, 1979; McKey, 1979; Rhoades, 1979; Coley *et al.*, 1985; Givnish, 1986; Gulmon and Mooney, 1986; Bazzaz *et al.*, 1987; Fagerstrom, 1989). This might be termed the adaptationist tradition (after Gould and Lewontin, 1979), in that the observed characters of plants are assumed to represent an optimal adaptive compromise between conflicting selective demands.

However, it is not clear that all traits conferring resistance must necessarily be subject to selective trade-offs. The main theoretical alternative to constraints imposed by selection are constraints on evolution caused by the structure of genetic variation within populations, which may be affected by factors like plant mating system, genetic drift and gene flow. For example, even when improved resistance is consistently favoured by directional selection, this trait will still fail to evolve if suitable genetic variation is lacking within local populations. For want of a better term, the analysis of these alternative factors will be called here the 'evolutionary genetics' approach to understanding constraints.

12.4 The adaptationist approach

If plants are not maximally resistant, the first response of a strict adaptationist would be to assume that this is because natural selection has not favoured this outcome. The primary research strategy, according to this paradigm, is to search for conflicting selective demands that interfere with evolution of maximal resistance. In other words, traits conferring resistance may have associated fitness costs.

Several distinct types of competing selective demands have been identified that may potentially prevent an increase in resistance. First, many authors have emphasized that allocation of scarce resources (energy or nutrients) to defence precludes their use for other essential functions, such as growth or reproduction (e.g. Gulmon and Mooney, 1986; Bazzaz *et al.*, 1987; Fagerstrom, 1989). As a result, individuals that maximize resistance are likely to have lower lifetime fitness than those balancing resource allocation more equitably between defence and other key activities. Investment in defence conflicts not only with maximization of current growth, but also involves an opportunity cost, in that diversion of resources to defence rather than to the production of additional acquisition structures (roots, leaves) necessarily limits future growth (Givnish, 1986; Gulmon and Mooney, 1986). In this sense, selection for an optimal resource allocation to defence versus growth is a special case of the general problem of timing allocation decisions in life history evolution (Williams, 1966).

However, it is important to recognize that not all fitness costs associated with parasite resistance can be understood in terms of resource allocation trade-offs. A prime example is discussed by Alexander (1989) who found that male plants of *Silene alba* that had higher rates of flower production suffered more infection by the pathogen, *Ustilago violacea*. In this case, plants can avoid exposure to pathogens only by producing fewer flowers, which directly lowers their reproductive success. Thus, resistance is costly here because a plant trait that improves fitness in a pathogen-free environment is precisely what makes plants vulnerable when pathogens are present.

Another type of selective conflict that is independent of resource allocation trade-offs concerns cases where traits conferring resistance interfere with plant tolerance of abiotic environmental stresses. The cyanogenesis polymorphism in *Trifolium repens* is perhaps the clearest example. Genotypes capable of cyanide release are less palatable than non-releasers to a variety of herbivores (Kakes, 1989; Hughes, 1991), but are infrequent in colder habitats such as high latitude and high altitude sites, apparently because they suffer autotoxicity upon freezing damage to leaf tissues (Daday, 1965). This example illustrates that the form of selection on resistance traits may often vary greatly in different habitats. Thus, a complete analysis of costs will always be a challenging problem for experimentalists.

A final type of selective conflict involves biotic trade-offs. Traits that confer resistance to one parasite can also make a plant more vulnerable to other parasite species (reviewed by Linhart, 1991). Virtually all plant populations are simultaneously exploited by several parasite taxa. Thus, there may be wide variation in the form of natural selection on particular resistance traits, depending on which parasite taxa are most abundant or most damaging at any point in time (Maddox and Root, 1990; Marquis, 1990).

To summarize, there are many types of ecological factors that may conflict with selection for maximal parasite resistance in particular cases. Before considering whether selective conflicts or fitness costs are an inherent aspect of all resistance traits, some methodological issues regarding how such costs can be experimentally quantified will be discussed.

12.4.1 *Costs: methodological issues*

Parasite resistance can be viewed as costly if resistant hosts have lower fitness than susceptible hosts when both develop in an environment free from parasite attack. I have so far discussed this problem in terms of a conflict between different forces of selection. However, this conflict will exist only if certain organism phenotypes are impossible to achieve, that is, if there is no mechanism to create a phenotype that combines maximum parasite resistance with high fitness in other respects. Constraints on phenotype can arise either from pleiotropy or from close linkage between alleles affecting resistance and other traits. If resistance alleles have pleiotropic effects on other selected traits, or are tightly linked to alleles influencing other traits, then natural selection on these other traits can interfere with the evolution of improved resistance.

There are two basic approaches to measuring whether a genetic association exists between resistance and other fitness components, depending on whether resistance is controlled by a single locus or is considered to be a polygenic trait. For polygenically inherited resistance, Simms and Rausher (1987, 1989) advocated that costs be quantified by the additive genetic correlation between resistance and lifetime plant fitness in a parasite-free environment. This can be measured either by breeding experiments (e.g. Simms and Rausher, 1989), or by looking for a correlated response to selection on one trait in an artificial selection experiment (Falconer, 1981). For resistance controlled by alternative alleles at one or a small number of loci, costs can be analysed simply by comparing performance of identified genotypes in a parasite-free environment (e.g. Burdon and Muller, 1987; Kakes, 1989; Jorgensen and Jensen, 1990; Parker, 1990).

Several caveats must be kept in mind in evaluating studies of association between resistance and other plant fitness components. First, the distinction between phenotypic and genetic correlations has not always been made in previous work. Evolutionary arguments about costs of resistance require a knowledge of genetic correlations, yet these cannot be directly inferred from the raw phenotypic correlation between resistance and other traits (Falconer, 1981; Simms, 1992). To note one example, several authors have cited the study of Coley (1986) to support the contention that anti-herbivore traits are costly to plants. This study reported a significant negative correlation between leaf tannin content (a chemical defense against herbivory) and plant growth rate. However, the experimental design involved plants of unknown genetic relationship that had not experienced a common environment throughout life, so it does not allow partitioning of genetic versus non-genetic sources of co-variation. Thus, while the results are suggestive, they do not provide unambiguous evidence for a cost of resistance.

A second potential bias can occur if certain agents of evolution have acted on the population prior to sampling of plant genotypes. Transient trait correlations may be produced by genetic drift or migration (Hedrick *et al.*, 1978), or by 'correlational' natural selection within a generation that creates a phenotypic association between two characters (Endler, 1986; Phillips and Arnold, 1989). In either case, a spurious correlation between resistance and other traits may be found in any population of established plants that has been subject to these agents of evolution at earlier life history stages. In quantitative genetics experiments, this problem can be minimized by experimentally crossing individuals to obtain an unbiased sample of progeny genotypes (Berenbaum *et al.*, 1986; Mitchell-Olds and Rutledge, 1986; Simms and Rausher, 1989). Without experimental crosses, samples of established plants from nature provide only equivocal data about trade-offs (e.g. Cates, 1975; Coley, 1986; Bjorkman and Anderson, 1990).

The problem of non-random sampling is particularly acute in plant taxa whose mating systems limit genetic recombination (e.g. species with extensive self-pollination or clonal reproduction) because, in the absence of recombination, factors that generate trait correlations can have a cumulative effect over time (Hedrick *et al.*, 1978). In plant taxa with restricted recombination, many ecologically important traits are often found to covary with resistance polymorphisms (Parker, 1992). This might

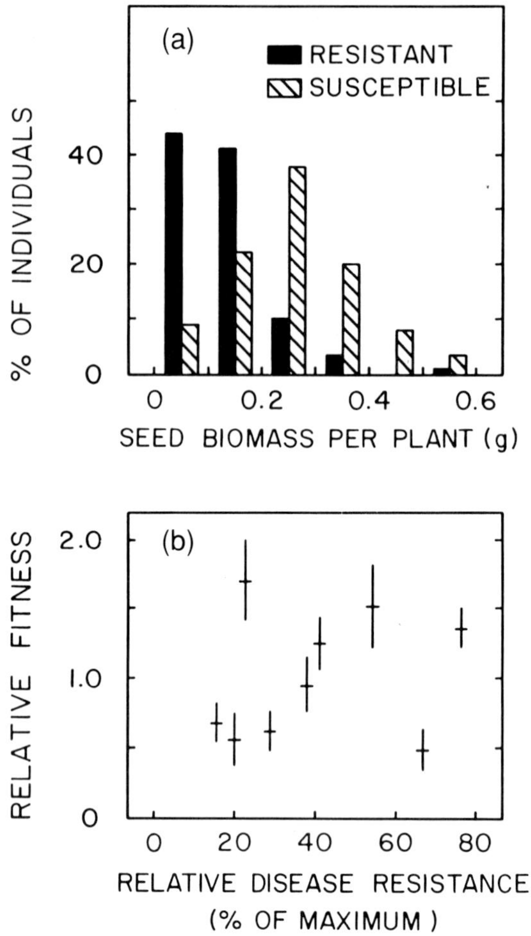

Figure 12.2. *(a) Lifetime reproductive success for resistant and susceptible genotypes of* A. bracteata, *when grown in a disease-free common garden. Plants were scored for resistance to* S. decipiens *in laboratory inoculation tests (Parker, 1991a), and then lifetime seed biomass was measured for 90 selfed progeny from each type. (b) Relation between disease resistance and fitness (± 1 s.e.) in a pathogen-free environment for recombinant progeny of genotypes shown in (a). Nine plant families derived from separate F2 individuals were each subdivided into two groups. For each family, one group was scored for disease resistance (Parker, 1991a), and then lifetime seed biomass was measured for a separate set of 18 individuals in a common garden. Lifetime seed biomass per plant was divided by the population mean for a measure of relative fitness. Disease resistance was scaled by dividing each family's mean by that of the resistant parent, which had the maximum resistance observed in this sample of plants (Parker, 1991a).*

suggest to some that there are trade-offs between parasite resistance and other key traits. However, the spurious nature of these trait associations can be illustrated with the following example. A population of the highly self-pollinated legume, *A. bracteata*, growing near Chicago, Illinois, was found to be dominated by two divergent lineages that differed strongly in disease resistance to local pathogens (Parker, 1991a). When these lineages were grown in a disease-free common garden, they differed strikingly in overall reproductive success (Figure 12.2a). Plants of the susceptible lineage achieved two-fold higher fecundity on average. If this study had considered only data from the genotypes that predominate in nature, these results might have been interpreted as evidence for a trade-off between disease resistance and overall plant fitness. However, these genotypes were then experimentally hybridized. Among recombinant progeny, disease resistance was completely uncorrelated with relative reproductive success in a pathogen-free environment (Figure 12.2b; $r_s = +0.17$; $P > 0.50$). Thus, the negative correlation between resistance and fitness among the dominant genotypes in nature (Figure 12.2a) cannot be viewed as evidence that resistance is costly, since this correlation could be altered by a single episode of outcrossing.

A final methodological problem in analysing costs concerns the choice of environment for comparing plant performance. In principle, the degree of fitness difference between resistant and susceptible phenotypes may be strongly affected by overall environmental quality, the intensity of competition and other environmental factors. This is particularly likely for cases where costs arise from resource allocation trade-offs, since a change in supply of a limiting resource should alter costs associated with investment in defence. For example, an increase in soil nitrogen concentration may allow plants to invest heavily in nitrogen-based defences such as alkaloids without compromising other metabolic needs (Johnson *et al.*, 1987). Thus, experiments to analyse costs should ideally include a range of different environmental conditions, and should focus especially on environments similar to those experienced by natural populations. In some cases, the apparent magnitude of costs has been totally altered by a change in growing conditions (Burdon and Muller, 1987). Yet in other systems, costs have been found to be insensitive to major changes in plant environment (Parker, 1990).

In summary, although trade-offs between parasite resistance and other plant fitness components are often discussed as a primarily physiological or ecological matter, careful genetic analyses are essential to analyse fully these costs.

12.4.2 *Are costs universal?*

Explicitly or implicitly, it has been widely assumed that traits conferring resistance to parasites must have fitness costs, so that evolution of defence is best understood in terms of the selective balance between costs and benefits (Whittaker and Feeny, 1971; Cates, 1975; McKey, 1979; Rhoades, 1979; Coley *et al.*, 1985; Givnish, 1986; Bazzaz *et al.*, 1987). But what is the justification for this assumption?

One widely cited form of support for this premise is the apparent loss of resistance that has occurred in plants following their isolation from particular parasites upon transport to an island or another continent (Janzen, 1975; Van der Plank, 1975; Harlan, 1976; Chew and Rodman, 1979). Diminished resistance after a period of isolation from a pest has been interpreted as evidence for active selection against resistance traits when they become unnecessary. However, there have been no studies actually measuring changes in resistance in particular plant lineages over time, so evidence for this purported evolutionary trend is scant. Moreover, even if such a pattern was found, it is difficult to rule out non-selective processes such as genetic drift and mutation pressure, which tend to eliminate any trait upon relaxation of selection for its maintenance (Kimura, 1983). Also, there are cases known where separation has apparently produced the opposite outcome, with plants displaying greater resistance to allopatric parasites (Harris, 1975; Elmhirst and Heath, 1987).

Several arguments support the contention that costs need not be a general attribute of parasite resistance in plants. First, as noted above for cyanogenesis in *Trifolium*, a fitness cost may exist only within a subset of the environments that a population occupies. Thus, even though resistance mechanisms of this type can potentially limit the ecological amplitude of a species, costs may be negligible within some habitats. Second, many traits that contribute to resistance also have other functions for the plant (Bazzaz *et al.*, 1987). If the optimum trait value for an alternative function is more extreme than that required for defence against parasites, then resistance will be achieved effectively without cost. For example, many leaves have a waxy surface layer that serves both to limit desiccation and to provide a barrier to invasion by microbes. If plants cannot survive abiotic climate stresses without substantial investment in wax production, then allocation of resources for this function cannot be viewed as a cost of parasite resistance, even if effective resistance is in fact an outcome of this investment. Another possible example is the trait of rhizome bud scales in the herbaceous perennial, *Podophyllum peltatum* (Berberidaceae). During shoot emergence, these expand above the soil surface and form a sheath to protect delicate young leaves from frost damage and abrasion by soil. However, bud scales are also critical in limiting infection by the specialist rust, *Puccinia podophylli* (Parker, 1988). Here again, a cost of resistance would exist only if it could be shown that defence against pathogens required a more extreme phenotype than that needed for the other functions of this trait.

For many plant–pathogen interactions, a third reason why costs might sometimes be trivial is that resistance is essentially a problem of parasite recognition. Inducible plant defence mechanisms, such as localized cell death in response to invasion, are thought to be ubiquitous traits that appeared early in the evolution of higher plants (Heath, 1991). In order for a pathogen to establish a successful parasitic relationship, it must invade host tissues and acquire nutrients without signalling its presence in a way that would trigger these active host defence mechanisms. Thus, plants can achieve resistance simply by developing the ability to recognize any one of several different aspects of a pathogen's biochemical phenotype. Because host recognition of a pathogen can occur by multiple alternative pathways, it seems likely that many of the genetic changes conferring recognition ability will have negligible fitness costs for the host. This conclusion is supported by several studies of single-locus disease resistance mechanisms, which have found that resistance genes generally have no harmful effect on plant performance in a pathogen-free environment (Brinkman and Frey, 1977; Burdon and Muller, 1987; Jorgensen and Jensen, 1990; Parker, 1990).

A final argument as to why fitness costs of resistance may not be significant is that such costs are subject to evolutionary modification (Simms, 1992). For example, Lenski (1988) found that mutations in *Escherichia coli* conferring resistance to a bacteriophage had a significant cost associated, in that they reduced competitive fitness. However, after evolution of resistant strains in the absence of pathogens for 400 generations, modifier alleles at other loci had evolved that reduced the detrimental effect of these resistance mutations. To date, similar studies of modifier evolution have not been performed in plants. However, it has long been known that the effect of a resistance gene on fitness may vary considerably in different genetic backgrounds (Frey and Browning, 1971). This supports the idea that costs are not necessarily a fixed attribute of a resistance gene, but can be mitigated by evolution at other loci.

It is not implied that all parasite resistance traits in plants are free of fitness costs.

Clearly, estimates of biochemical synthesis costs indicate that significant carbon investment is required for particular defensive chemicals (Chew and Rodman, 1979; but see Mihaliak *et al.*, 1991 for a caveat regarding how such costs might be overestimated). Morever, many resistance traits are inducible, produced only in direct response to parasite damage (Karban and Myers, 1989). It is difficult to explain why any trait conferring resistance would not be expressed constitutively, unless there is a significant cost in having it when not needed (Harvell, 1990). This inference is supported by the study of Baldwin *et al.* (1990), which showed that plants induced to synthesize a chemical defence suffered a significant loss of fecundity, relative to similarly treated plants that had been manipulated to minimize defence investment. In conclusion, the studies reviewed here demonstrate that neither the presence nor the absence of major costs can be assumed *a priori*. Careful analysis of each particular system is needed to establish whether costs exist and, especially, to define the ecological circumstances that determine their magnitude.

12.5 Non-adaptive explanations for constraint

The working assumption of adaptationists is that the current state of a character represents the best compromise between the multiple selective demands experienced by a population. Implicit in this view is the notion that populations are at or near an evolutionary equilibrium. If we discard this premise, it is clear that the most basic reason why a population may fail to have an optimal level of resistance is that there has simply been insufficient time for it to reach a selective equilibrium with existing parasite populations. Two distinct processes may lead to such a time-lag constraint. First, ecological processes such as parasite species turnover within local communities due to invasion and extinction (Anderson and May, 1986), and temporal fluctuation in relative abundance of different parasites (e.g. Marquis, 1990), mean that all plant populations are likely to experience a continually changing spectrum of selective regimes imposed by natural enemies. A second form of time lag can arise from a difference in microevolutionary rates between host and parasite. If it is assumed that most parasites have much shorter generation times than their hosts and are therefore capable of more rapid evolution (see below), then parasite counteradaptation will erode existing host defences faster than they can be restored by evolution in the host. In either case, current host defence phenotypes may often deviate from the selective optimum.

A case in point concerns the American chestnut, *Castanea dentata*. Before 1900, this tree was ecologically dominant over a broad region of eastern North America, and it had therefore presumably adapted effectively to resist parasites endemic to this area. Yet it was essentially wiped out over its entire geographic range following the accidental introduction of the chestnut blight pathogen, *Cryphonectria parasitica*, early in this century (Hepting, 1974). It now persists only as relict populations of stump sprouts. These are clonal offshoots of trees destroyed by blight and thus retain the susceptible characteristics of their parents. Since sprouts are typically killed by blight infection before they have had much opportunity for sexual reproduction, the recruitment of new genotypes into chestnut populations is extremely restricted at present.

Thus, despite the catastrophic impact of blight and the opportunity for intense natural selection favouring resistance, it is uncertain whether these populations will ever evolve resistance to this parasite.

A second category of non-adaptive constraints involves forces that restrict the local diversity of genotypes on which selection can act. Severe population bottlenecks or founder events in colonizing species may result in the loss of genetic variation and therefore limit the future evolutionary potential of hosts to adapt to parasite attack (O'Brien and Evermann, 1988). Also, constraints on gene flow due to limited mobility or population fragmentation may isolate local host populations from sources of new alleles that may be critical in developing resistance. An example is where all genotypes in an isolated population of *A. bracteata* were found to be completely susceptible to local strains of *S. decipiens* (Parker, unpublished). Yet plants sampled from a population 1 km away were highly resistant to these same pathogens. This plant has no apparent mechanisms for long distance seed dispersal, and gene flow via pollen is minimal owing to a high frequency of self-fertilization (Parker, 1991a,b). Thus, isolated populations of this species may often have maladaptive levels of disease resistance, as a result of the low rate of input of new resistance genes into local areas. Given the accelerating rate of population fragmentation for nearly all species caused by human activities (Hanski and Gilpin, 1991), this type of constraint may become increasingly important in the future.

A third category of constraints may be imposed by plant mating systems. The hypothesis that parasite attack is a key selective agent favouring outcrossed sexuality in plants has received considerable attention (reviewed in Hamilton *et al.*, 1990; Parker, 1992). If valid, this theory would imply that any plant whose mating system results in restricted genetic recombination should have impaired ability to adapt to parasites. Evolutionary reversal may be difficult for many traits associated with uniparental mating systems, such as cleistogamy in self-fertilizers, or odd-ploidy level in many apomicts (Stebbins, 1974; Grant, 1981; Bull and Charnov, 1985). Thus, species with limited recombination may suffer increased vulnerability, because they are likely to have only a limited capacity for alteration of reproductive behaviour in response to parasite attack. To evaluate this potential type of constraint, it is important to distinguish between two distinct processes by which a lack of sex could interfere with adaptation to parasites.

First, it is most often argued that sex is favoured because parasites can evolve very rapidly when they encounter a host population with a uniform defensive phenotype (Tooby, 1982; Hamilton *et al.*, 1990). If parasites can quickly counteradapt to particular host resistance mechanisms, then any host lineage that gives up sexual reproduction becomes an easy target, and should therefore accumulate a heavy burden of adapted enemies. By contrast, hosts that reproduce sexually offer parasites a diverse and continually changing set of different defence phenotypes. This may allow sexual populations to escape the full force of parasites that have adapted to host phenotypes prevalent in previous generations. To test this argument, we therefore need to ask: what is the evidence that parasites can always counteradapt to nullify a consistently deployed host defence? In agriculture, the ability of pathogens rapidly to overcome plant resistance genes is notorious (Stakman *et al.*, 1943; Barrett, 1981). But even under conditions of extensive host genetic uniformity in agriculture, there are many cases of durable resistance (Buddenhagen, 1977; Schlosser, 1980; Johnson, 1984). Why parasites have adapted at vastly different rates to various plant resistance traits remains an enigma (Gould, 1991). However, these examples suggest that not all resistance mechanisms deployed in uniform host populations are immediately vulnerable to pathogen counteradaptation.

Furthermore, in natural communities, there are many examples of hosts with restricted recombination that are abundant and ecologically successful (Allard, 1965). Among plants that rely on vegetative multiplication, numerous cases are known where individual clones have proliferated to immense size and have apparently survived for hundreds or thousands of years (Cook, 1985). This implies that a persistent, widely distributed genotype with a fixed defensive phenotype need not always be in extreme danger of elimination by parasites (Parker, 1988).

Moreover, asexual plant populations often contain moderate or high levels of genetic diversity (Ellstrand and Roose, 1987), which has two important implications. First, these populations do not necessarily lack the capacity for significant evolution in response to various types of natural selection (Lynch and Gabriel, 1983). Second, to the extent that genetic diversity among host individuals is an immediate factor limiting parasite impact, there may not be a qualitative difference in susceptibility between sexual and asexual populations. In fact, a number of studies have compared parasite abundance in experimental populations where the degree of genetic uniformity among hosts has been manipulated. In most of these studies, the level of genetic diversity among hosts had only a weak or inconsistent effect on parasite impact (Jarosz and Levy, 1988; Parker, 1989; Alexander, 1991; but see Powers, 1991).

To summarize, rapid pathogen adaptation has been widely touted as a fundamental problem that plants can counter only by sexual reproduction, to produce a diversity of defensive phenotypes in space and in time. However, observations on natural and experimental plant populations imply that the link between host genetic uniformity and parasite susceptibility may vary widely in different systems. A great deal more comparative and experimental work will thus be necessary before the importance of this type of constraint can be judged.

A distinct mechanism by which restriction of recombination can interfere with plant adaptation to parasites involves cases where there are non-random associations between resistance and other selected traits. Gametic disequilibrium between alleles at separate loci may be generated by several evolutionary processes (Hedrick *et al.*, 1978). In populations with restricted recombination, such multi-locus structure can severely limit how a population responds to different agents of selection (Parker, 1992). Non-random associations between resistance and other ecologically important traits have been observed in a number of plant species where recombination is restricted either by extensive self-fertilization (Muona *et al.*, 1984; Burdon and Muller, 1987), or by clonal reproduction (Harberd, 1961, 1967; Burdon *et al.*, 1980). If the force of natural selection on these correlated traits is stronger than that exerted by parasites, then resistance may evolve in a non-adaptive or even a maladaptive fashion.

Observations of a natural population of the highly self-pollinated *A. bracteata* provide an example (Parker, 1991b). Over a 2-year period, plant lineages that were

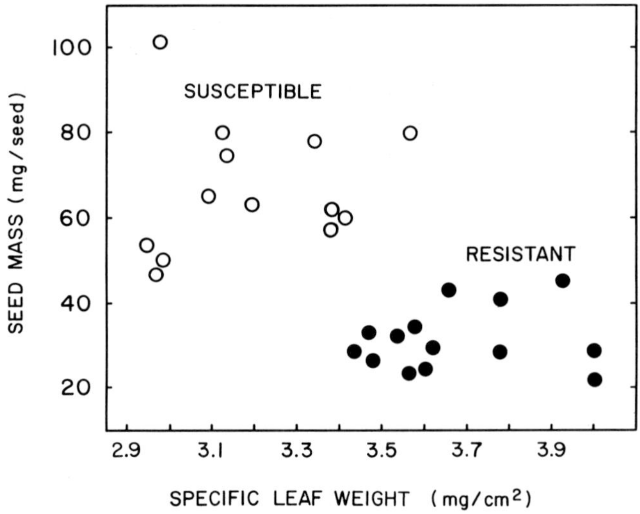

Figure 12.3. *Bivariate plot of morphological traits correlated with disease resistance polymorphism in* A. bracteata. *Plants were tested for resistance to* S. decipiens *(Parker, 1991b), and then selfed progeny of 13 susceptible and 14 resistant families were grown in a common garden (*$n = 18$ *plants per family). Family means for specific leaf weight and seed mass are plotted for resistant (●) and susceptible genotypes (○). Kruskal–Wallis tests indicated that resistant and susceptible families differed significantly for both traits (*$P < 0.001$*).*

resistant to *S. decipiens* were found to decrease significantly in abundance relative to susceptible lineages (resistant genotypes declined from 41 to 16% of the population). Analysis of plants in a disease-free common garden showed that numerous morphological and life history traits were correlated with the disease resistance polymorphism in this population. Figure 12.3 shows two examples. Plant lines that were susceptible to disease had leaves of lower specific leaf weight, and also produced larger seeds relative to resistant plants. Variation in these and other traits (see Parker, 1991b) may strongly affect relative plant success in response to light environment, drought stress, or competition (Schnee and Waller, 1986; Fitter and Hay, 1987). Because disease resistance was highly correlated with many other ecologically important traits, it is likely that natural selection on one or more of these traits was responsible for the observed loss of resistance from this population. This constraint on resistance arises directly from the restriction of recombination imposed by the plant's highly self-fertilized mating system. Selfing impedes the formation of lineages that combine resistance with other beneficial traits while, in a freely outcrossing population, optimal combinations of traits could be generated much more readily. Given that correlations between resistance and other ecologically important traits appear to be widespread in plant taxa with limited recombination, this is likely to be a major type of evolutionary constraint for plants with these mating systems.

To summarize, research on constraints within what I have called the 'evolutionary genetics' tradition has shown that numerous factors other than optimizing selection may cause plant populations to fail to achieve effective parasite resistance: time lags, population bottlenecks, constraints on gene flow, and finally, plant mating systems that limit free genetic recombination. Individually, each of these may or may not apply to any given plant species, but, collectively, these factors are likely to occur in a wide variety of natural populations. Thus, they cannot be ignored in any comprehensive theory about constraints.

12.6 Conclusions: reconciling the two research traditions

Historically, the two research traditions for analysing constraints have tended to draw practitioners from distinct disciplines, and have thus created bodies of work with very different strengths and biases. Typically, researchers working in the adaptationist tradition have had a functional orientation, seeking to understand the precise chemical or morphological mechanisms responsible for parasite resistance, and how various plant processes (physiological, biochemical, developmental) interact with the environment to cause variation in resistance among individuals. The power of the functional approach is obvious from the impressive depth of knowledge that we now have about mechanisms of parasite resistance in plants (e.g. Rosenthal and Janzen, 1979). Yet too often, this functional approach has been coupled with an exclusive focus on natural selection as the only significant agent of evolution, and has thus led to the bias that selective trade-offs are the only type of constraint on resistance that need be considered.

The evolutionary genetics tradition suffers from a complementary limitation. Many researchers with backgrounds in population genetics or quantitative genetics have limited expertise in the functional biology of parasite resistance, and thus attempt to account for constraints with only a superficial understanding of the actual mechanisms responsible for resistance in the organisms they study (e.g. Simms and Rausher, 1989; Parker, 1990). The strength of this approach is that alternative hypotheses about constraint are often explicitly stated and experimentally tested. However, it is obvious that accurate hypotheses about constraints can best be formulated if we have prior knowledge of the actual physiological and biochemical mechanisms responsible for resistance. On a fundamental level, these two disparate approaches are thus complementary rather than in conflict, and the most meaningful studies are likely to be those that bridge both traditions.

Is it possible to reach any broad conclusions about what types of constraints on parasite resistance are most pervasive or powerful in evolution? I think that we have not yet reached the stage where synthetic conclusions are possible. However, I offer one opinion. Viewed over long stretches of evolutionary time, it is likely that within any given plant lineage a number of alternative methods for achieving resistance will be tried out. Any trait that consistently increases resistance to a broad spectrum of parasites at minimal cost will be rapidly fixed. By contrast, traits that provide resistance only to some parasites while increasing susceptibility to others, or that interfere with other ecological demands on plants, will be subject to variable selection and are unlikely to become fixed. Thus, even if it is the case that most traits conferring resistance have minimal fitness costs, most of the variation in resistance that we see in present day populations is likely to involve traits that are subject to conflicting forces of selection. This may bias our perception as to the importance of selective conflicts acting on resistance, because ecologists and evolutionary biologists naturally tend to focus on individual variation.

While we have answers concerning the role of particular constraints for particular species, the general question of why plants have adapted unevenly to resist some parasites and not others remains a deep mystery. This problem is of fundamental interest from both a theoretical and applied point of view, and it thus deserves continued vigorous research attention.

References

Alexander, H.M. (1989) An experimental field study of anther-smut disease of *Silene alba* caused by *Ustilago violacea*: genotypic variation and disease incidence. *Evolution* **43**, 835–847.

Alexander, H.M. (1991) Plant population heterogeneity and pathogen and herbivore levels: a field experiment. *Oecologia* **86**, 125–131.

Allard, R.W. (1965) Genetic systems associated with colonizing ability in predominantly self-pollinated species. In: *The Genetics of Colonizing Species* (eds H.G. Baker and G.L. Stebbins). Academic Press, New York, pp. 49–75.

Anderson, R.M. and May, R.M. (1986) The invasion, persistence and spread of infectious diseases within animal and plant communities. *Phil. Trans. R. Soc. Lond. B* **314**, 533-570.

Baldwin, I.T., Sims, C.L. and Kean, S.E. (1990) The reproductive consequences associated with inducible alkaloidal responses in wild tobacco. *Ecology* **71**, 252–262.

Barrett, J.A. (1981) The evolutionary consequences of monoculture. In: *Genetic Consequences of Man Made Change* (eds J.A. Bishop and L.M. Cook). Academic Press, London, pp. 209–248.

Bazzaz, F.A., Chiariello, N.R., Coley, P.D. and Pitelka, L.F. (1987) Allocating resources to reproduction and defense. *Bioscience* **37**, 58–67.

Berenbaum, M.R., Zangerl, A.R. and Nitao, J.K. (1986) Constraints on chemical coevolution: wild parsnips and the parsnip webworm. *Evolution* **40**, 1215–1228.

Bjorkman, C. and Anderson, D.B. (1990) Trade-off among antiherbivore defences in a South American blackberry (*Rubus bogotensis*). *Oecologia* **85**, 247–249.

Brinkman, M.A. and Frey, K.J. (1977) Growth analysis of isoline-recurrent parent grain yield differences in oats. *Crop Sci.* **17**, 426–430.

Broglie, K., Chet, I., Holliday, M., Cressman, R., Biddle, P., Knowlton, S., Mauvais, C.J. and Broglie, R. (1991) Transgenic plants with enhanced resistance to the fungal pathogen *Rhizoctonia solani*. *Science* **254**, 1194–1197.

Buddenhagen, I.W. (1977) Resistance and vulnerability of tropical crops in relation to their evolution and breeding. *Ann. N.Y. Acad. Sci.* **287**, 309–326.

Bull, J.J. and Charnov, E.L. (1985) On irreversible evolution. *Evolution* **39**, 1149–1155.

Burdon, J.J., Marshall, D.R. and Groves, R.H. (1980) Isozyme variation in *Chondrilla juncea* in Australia. *Aust. J. Bot.* **28**, 193–198.

Burdon, J.J. and Muller, W.J. (1987) Measuring the cost of resistance to *Puccinia coronata* in *Avena fatua*. *J. Appl. Ecol.* **24**, 191–200.

Cates, R.G. (1975) The interface between slugs and wild ginger: some evolutionary aspects. *Ecology* **56**, 391–400.

Chew, F.S. and Rodman, J.E. (1979) Plant resources for chemical defense. In: *Herbivores: their*

Interaction with Secondary Plant Metabolites (eds G.A. Rosenthal and D.H. Janzen). Academic Press, New York, pp 271–307.

Coley, P.D., Bryant, J.P. and Chapin, F.S. (1985) Resource availability and plant anti-herbivore defense. *Science* **230**, 895–899.

Coley, P.D. (1986) Costs and benefits of defense by tannins in a neotropical tree. *Oecologia* **70**, 238–241.

Cook, R.E. (1985) Growth and development in clonal plant populations. In: *Population Biology and Evolution of Clonal Organisms* (eds J.B.C. Jackson, L.W. Buss and R.E. Cook). Yale University Press, New Haven, CT, pp. 259–296.

Daday, H. (1965) Gene frequencies in wild populations of *Trifolium repens*. IV. Mechanism of natural selection. *Heredity* **20**, 355–365.

Ellstrand, N.C. and Roose, M.L. (1987) Patterns of genotypic diversity in clonal plant species. *Am. J. Bot.* **74**, 123–131.

Elmhirst, J.F. and Heath, M.C. (1987) Interactions of the bean rust and cowpea rust fungi with species of the *Phaseolus–Vigna* plant complex. I. Fungal growth and development. *Can. J. Bot.* **65**, 1096–1107.

Endler, J.A. (1986) *Natural Selection in the Wild*. Princeton University Press, Princeton, New Jersey.

Fagerstrom, T. (1989) Anti-herbivore chemical defense in plants: a note on the concept of cost. *Am. Nat.* **133**, 281–287.

Falconer, D.S. (1981) *Introduction to Quantitative Genetics*. Second edition. Longman, London.

Fitter, A.H. and Hay, R.K.M. (1987) *Environmental Physiology of Plants*. Second edition. Academic Press, London.

Frey, K.J. and Browning, J.A. (1971) Association between genetic factors for crown rust resistance and yield in oats. *Crop Sci.* **11**, 757–760.

Givnish, T.J. (1986) Economics of biotic interactions. In: *On the Economy of Plant Form and Function* (ed. T.J. Givnish). Cambridge University Press, Cambridge, pp. 667–680.

Gould, F. (1991) The Evolutionary Potential of Crop Pests. *Am. Sci.* **79**, 496–507.

Gould, S.J. and Lewontin, R.C. (1979) The spandrels of San Marco and the Panglossian paradigm: a critique of the adaptationist programme. *Proc. R. Soc. Lond. B* **205**, 581–598.

Grant, V. (1981) *Plant Speciation*. Second edition. Columbia University Press, New York.

Gulmon, S.L. and Mooney, H.A. (1986) Costs of defense and their effects on plant productivity. In: *On the Economy of Plant Form and Function* (ed. T.J. Givnish). Cambridge University Press, Cambridge, pp. 681–698.

Hamilton, W.D., Axelrod, R. and Tanese, R. (1990) Sexual reproduction as an adaptation to resist parasites (a review). *Proc. Natl. Acad. Sci. USA* **87**, 3566–3573.

Hanski, I. and Gilpin, M. (1991) Metapopulation dynamics: brief history and conceptual domain. *Biol. J. Linn. Soc.* **42**, 3–16.

Harberd, D.J. (1961) Observations on population structure and longevity of *Festuca rubra*. *New Phytol.* **60**, 184–206.

Harberd, D.J. (1967) Observations on natural clones in *Holcus mollis*. *New Phytol.* **66**, 401–408.

Harlan, J.R. (1976) Diseases as a factor in plant evolution. *Ann. Rev. Phytopathol.* **14**, 31-51.

Harris, M.K. (1975) Allopatric resistance: searching for sources of insect resistance for use in agriculture. *Environ. Entomol.* **4**, 661–669.

Harvell, C.D. (1990) The ecology and evolution of inducible defenses. *Q. Rev. Biol.* **65**, 323–340.

Heath, M.C. (1991) Evolution of resistance to fungal parasitism in natural ecosystems. *New Phytol.* **119**, 331–343.

Hedrick, P., Jain, S. and Holden, L. (1978) Multilocus systems in evolution. *Evol. Biol.* **11**, 101–184.

Hepting, G.H. (1974) Death of the American chestnut. *J. Forest History* **18**, 60–67.

Hughes, M.A. (1991) The cyanogenic polymorphism in *Trifolium repens* (white clover). *Heredity* **66**, 105–115.

Janzen, D.H. (1975) Behavior of *Hymenaea coubaril* when its predispersal seed predator is absent. *Science* **189**, 145–147.

Jarosz, A.M. and Levy, M. (1988) Effects of habitat and population structure on powdery mildew epidemics in experimental *Phlox* populations. *Phytopathol.* **78**, 358–362.

Johnson, N.D., Liu, B. and Bentley, B.L. (1987) The effects of nitrogen fixation, soil nitrate, and defoliation on the growth, alkaloids, and nitrogen levels of *Lupinus succulentus* (Fabaceae). *Oecologia* **74**, 425–431.

Johnson, R. (1984) A critical analysis of durable resistance. *Ann. Rev. Phytopathol.* **22**, 309–330.

Jorgensen, J.H. and Jensen, H.P. (1990) Effects of 'unnecessary' powdery mildew resistance genes on agronomic properties of spring barley. *Norsk Landbruksforsking Suppl.* 125–130.

Kakes, P. (1989) An analysis of the costs and benefits of the cyanogenic system in *Trifolium repens*. *Theor. Appl. Genet.* **77**, 111–118.

Karban, R. and Myers, J.H. (1989) Induced plant responses to herbivory. *Ann. Rev. Ecol. Syst.* **20**, 331–348.

Kimura, M. (1983) *The Neutral Theory of Molecular Evolution.* Cambridge University Press, Cambridge.

Lenski, R.E. (1988) Experimental studies of pleiotropy and epistasis in *Escherichia coli*. II. Compensation for maladaptive effects associated with resistance to virus T4. *Evolution* **42**, 433–440.

Linhart, Y.B. (1991) Disease, parasitism and herbivory: multidimensional challenges in plant evolution. *Trends Ecol. Evol.* **6**, 392–396.

Lynch, M. and Gabriel, W. (1983) Phenotypic evolution and parthenogenesis. *Am. Nat.* **122**, 745–764.

McKey, D. (1979) The distribution of secondary compounds within plants. In: *Herbivores: Their Interaction with Secondary Plant Metabolites* (eds G.A. Rosenthal and D.H. Janzen). Academic Press, New York, pp. 55–133.

Maddox, G.D. and Root, R.B. (1990) Structure of the encounter between goldenrod (*Solidago altissima*) and its diverse insect fauna. *Ecology* **71**, 2115–2124.

Marquis, R.J. (1990) Genotypic variation in leaf damage in *Piper arieianum* (Piperaceae) by a multispecies assemblage of herbivores. *Evolution* **44**, 104–120.

Mihaliak, C.A., Gershenzon, J. and Croteau, R. (1991) Lack of rapid monoterpene turnover in rooted plants: implications for theories of plant chemical defense. *Oecologia* **87**, 373–376.

Mitchell-Olds, T. and Rutledge, J.J. (1986) Quantitative genetics in natural plant populations: a review of the theory. *Am. Nat.* **127**, 379–402.

Muona, O., Allard, R.W. and Webster, R.K. (1984) Evolution of disease resistance and quantitative characters in barley Composite Cross II: independent or correlated? *Hereditas* **101**, 143–148.

O'Brien, S.J. and Evermann, J.F. (1988) Interactive influence of infectious disease and genetic diversity in natural populations. *Trends Ecol. Evol.* **3**, 254–259.

Parker, M.A. (1988) Genetic uniformity and disease resistance in a clonal plant. *Am. Nat.* **132**, 538–549.

Parker, M.A. (1989) Disease impact and local genetic diversity in the clonal plant *Podophyllum peltatum*. *Evolution* **43**, 540–547.

Parker, M.A. (1990) The pleiotropy theory for polymorphism of disease resistance genes in plants. *Evolution* **44**, 1872–1875.

Parker, M.A. (1991a) Local genetic differentiation for disease resistance in a selfing annual. *Biol. J. Linn. Soc.* **42**, 337–349.

Parker, M.A. (1991b) Non-adaptive evolution of disease resistance in an annual legume. *Evolution* **45**, 1209–1217.

Parker, M.A. (1992) Disease and plant population genetic structure. In: *Plant Resistance to Herbivores and Pathogens* (eds R.S. Fritz and E.L. Simms). University of Chicago Press, Chicago, pp. 345–362

Phillips, P.C. and Arnold, S.J. (1989) Visualizing multivariate selection. *Evolution* **43**, 1209–1222.

Powers, A.G. (1991) Virus spread and vector dynamics in genetically diverse plant populations. *Ecology* **72**, 232–241.

Rhoades, D.F. (1979) Evolution of plant chemical defense against herbivores. In: *Herbivores: their Interaction with Secondary Plant Metabolites* (eds G.A. Rosenthal and D.H. Janzen). Academic Press, New York, pp 3–54.

Rosenthal, G.A. and Janzen, D.H. (1979) *Herbivores: their Interaction with Secondary Plant Metabolites.* Academic Press, New York.

Schlosser, E.W. (1980) Preformed internal chemical defenses. In: *Plant Disease,* Volume 5 (eds J.G. Horsfall and E.B. Cowling). Academic Press, New York, pp. 161–177.

Schnee, B.K. and Waller, D.M. (1986) Reproductive behavior of *Amphicarpaea bracteata*, (Leguminosae), an amphicarpic annual. *Am. J. Bot.* **73**, 376–386.

Simms, E.L. and Rausher, M.D. (1987) Costs and benefits of plant resistance to herbivory. *Am. Nat.* **130**, 570–581.

Simms, E.L. and Rausher, M.D. (1989) The evolution of resistance to herbivory in *Ipomoea purpurea*. II. Natural selection by insects and costs of resistance. *Evolution* **43**, 573–585.

Simms, E.L. (1992) Costs of plant resistance to herbivores. In: *Plant Resistance to Herbivores and Pathogens* (eds R.S. Fritz and E.L. Simms). University of Chicago Press, Chicago, pp. 392–425.

Stakman, E.C., Loegering, W.Q., Cassell, R.C. and Hines, L. (1943) Population trends of physiological races of *Puccinia tritici* in the United States for the period 1930 to 1941. *Phytopathol.* **33**, 884–898.

Stebbins, G.L. (1974) *Flowering Plants: Evolution above the Species Level.* Harvard University Press, Cambridge, Massachusetts.

Tooby, J. (1982) Pathogens, polymorphism, and the evolution of sex. *J. Theor. Biol.* **97**, 557–576.

Van der Plank, J.E. (1975) *Principles of Plant Infection.* Academic Press, New York.

Whittaker, R.H. and Feeny, P.P. (1971) Allelochemics: chemical interactions between species. *Science* **171**, 757–770.

Williams, G.C. (1966) Natural selection, the costs of reproduction, and a refinement of Lack's principle. *Am. Nat.* **100**, 687–690.

Index